D1262082

volume 104

[lectur]e notes in pure and applied mathematics

[lin]ear systems over [c]ommutative rings

James W. Brewer
John W. Bunce
F. S. Van Vleck

LINEAR SYSTEMS OVER COMMUTATIVE RINGS

LECTURE NOTES

IN PURE AND APPLIED MATHEMATICS

1. *N. Jacobson,* Exceptional Lie Algebras
2. *L. -Å. Lindahl and F. Poulsen,* Thin Sets in Harmonic Analysis
3. *I. Satake,* Classification Theory of Semi-Simple Algebraic Groups
4. *F. Hirzebruch, W. D. Newmann, and S. S. Koh,* Differentiable Manifolds and Quadratic Forms (out of print)
5. *I. Chavel,* Riemannian Symmetric Spaces of Rank One (out of print)
6. *R. B. Burckel,* Characterization of C(X) Among Its Subalgebras
7. *B. R. McDonald, A. R. Magid, and K. C. Smith,* Ring Theory: Proceedings of the Oklahoma Conference
8. *Y.-T. Siu,* Techniques of Extension on Analytic Objects
9. *S. R. Caradus, W. E. Pfaffenberger, and B. Yood,* Calkin Algebras and Algebras of Operators on Banach Spaces
10. *E. O. Roxin, P.-T. Liu, and R. L. Sternberg,* Differential Games and Control Theory
11. *M. Orzech and C. Small,* The Brauer Group of Commutative Rings
12. *S. Thomeier,* Topology and Its Applications
13. *J. M. Lopez and K. A. Ross,* Sidon Sets
14. *W. W. Comfort and S. Negrepontis,* Continuous Pseudometrics
15. *K. McKennon and J. M. Robertson,* Locally Convex Spaces
16. *M. Carmeli and S. Malin,* Representations of the Rotation and Lorentz Groups: An Introduction
17. *G. B. Seligman,* Rational Methods in Lie Algebras
18. *D. G. de Figueiredo,* Functional Analysis: Proceedings of the Brazilian Mathematical Society Symposium
19. *L. Cesari, R. Kannan, and J. D. Schuur,* Nonlinear Functional Analysis and Differential Equations: Proceedings of the Michigan State University Conference
20. *J. J. Schäffer,* Geometry of Spheres in Normed Spaces
21. *K. Yano and M. Kon,* Anti-Invariant Submanifolds
22. *W. V. Vasconcelos,* The Rings of Dimension Two
23. *R. E. Chandler,* Hausdorff Compactifications
24. *S. P. Franklin and B. V. S. Thomas,* Topology: Proceedings of the Memphis State University Conference
25. *S. K. Jain,* Ring Theory: Proceedings of the Ohio University Conference
26. *B. R. McDonald and R. A. Morris,* Ring Theory II: Proceedings of the Second Oklahoma Conference
27. *R. B. Mura and A. Rhemtulla,* Orderable Groups
28. *J. R. Graef,* Stability of Dynamical Systems: Theory and Applications
29. *H.-C. Wang,* Homogeneous Branch Algebras
30. *E. O. Roxin, P.-T. Liu, and R. L. Sternberg,* Differential Games and Control Theory II
31. *R. D. Porter,* Introduction to Fibre Bundles
32. *M. Altman,* Contractors and Contractor Directions Theory and Applications
33. *J. S. Golan,* Decomposition and Dimension in Module Categories
34. *G. Fairweather,* Finite Element Galerkin Methods for Differential Equations
35. *J. D. Sally,* Numbers of Generators of Ideals in Local Rings
36. *S. S. Miller,* Complex Analysis: Proceedings of the S.U.N.Y. Brockport Conference
37. *R. Gordon,* Representation Theory of Algebras: Proceedings of the Philadelphia Conference
38. *M. Goto and F. D. Grosshans,* Semisimple Lie Algebras
39. *A. I. Arruda, N. C. A. da Costa, and R. Chuaqui,* Mathematical Logic: Proceedings of the First Brazilian Conference

40. *F. Van Oystaeyen,* Ring Theory: Proceedings of the 1977 Antwerp Conference
41. *F. Van Oystaeyen and A. Verschoren,* Reflectors and Localization: Application to Sheaf Theory
42. *M. Satyanarayana,* Positively Ordered Semigroups
43. *D. L. Russell,* Mathematics of Finite-Dimensional Control Systems
44. *P.-T. Liu and E. Roxin,* Differential Games and Control Theory III: Proceedings of the Third Kingston Conference, Part A
45. *A. Geramita and J. Seberry,* Orthogonal Designs: Quadratic Forms and Hadamard Matrices
46. *J. Cigler, V. Losert, and P. Michor,* Banach Modules and Functors on Categories of Banach Spaces
47. *P.-T. Liu and J. G. Sutinen,* Control Theory in Mathematical Economics: Proceedings of the Third Kingston Conference, Part B
48. *C. Byrnes,* Partial Differential Equations and Geometry
49. *G. Klambauer,* Problems and Propositions in Analysis
50. *J. Knopfmacher,* Analytic Arithmetic of Algebraic Function Fields
51. *F. Van Oystaeyen,* Ring Theory: Proceedings of the 1978 Antwerp Conference
52. *B. Kedem,* Binary Time Series
53. *J. Barros-Neto and R. A. Artino,* Hypoelliptic Boundary-Value Problems
54. *R. L. Sternberg, A. J. Kalinowski, and J. S. Papadakis,* Nonlinear Partial Differential Equations in Engineering and Applied Science
55. *B. R. McDonald,* Ring Theory and Algebra III: Proceedings of the Third Oklahoma Conference
56. *J. S. Golan,* Structure Sheaves over a Noncommutative Ring
57. *T. V. Narayana, J. G. Williams, and R. M. Mathsen,* Combinatorics, Representation Theory and Statistical Methods in Groups: YOUNG DAY Proceedings
58. *T. A. Burton,* Modeling and Differential Equations in Biology
59. *K. H. Kim and F. W. Roush,* Introduction to Mathematical Consensus Theory
60. *J. Banas and K. Goebel,* Measures of Noncompactness in Banach Spaces
61. *O. A. Nielson,* Direct Integral Theory
62. *J. E. Smith, G. O. Kenny, and R. N. Ball,* Ordered Groups: Proceedings of the Boise State Conference
63. *J. Cronin,* Mathematics of Cell Electrophysiology
64. *J. W. Brewer,* Power Series Over Commutative Rings
65. *P. K. Kamthan and M. Gupta,* Sequence Spaces and Series
66. *T. G. McLaughlin,* Regressive Sets and the Theory of Isols
67. *T. L. Herdman, S. M. Rankin, III, and H. W. Stech,* Integral and Functional Differential Equations
68. *R. Draper,* Commutative Algebra: Analytic Methods
69. *W. G. McKay and J. Patera,* Tables of Dimensions, Indices, and Branching Rules for Representations of Simple Lie Algebras
70. *R. L. Devaney and Z. H. Nitecki,* Classical Mechanics and Dynamical Systems
71. *J. Van Geel,* Places and Valuations in Noncommutative Ring Theory
72. *C. Faith,* Injective Modules and Injective Quotient Rings
73. *A. Fiacco,* Mathematical Programming with Data Perturbations I
74. *P. Schultz, C. Praeger, and R. Sullivan,* Algebraic Structures and Applications Proceedings of the First Western Australian Conference on Algebra
75. *L. Bican, T. Kepka, and P. Nemec,* Rings, Modules, and Preradicals
76. *D. C. Kay and M. Breen,* Convexity and Related Combinatorial Geometry: Proceedings of the Second University of Oklahoma Conference
77. *P. Fletcher and W. F. Lindgren,* Quasi-Uniform Spaces
78. *C.-C. Yang,* Factorization Theory of Meromorphic Functions
79. *O. Taussky,* Ternary Quadratic Forms and Norms
80. *S. P. Singh and J. H. Burry,* Nonlinear Analysis and Applications
81. *K. B. Hannsgen, T. L. Herdman, H. W. Stech, and R. L. Wheeler,* Volterra and Functional Differential Equations

82. *N. L. Johnson, M. J. Kallaher, and C. T. Long,* Finite Geometries: Proceedings of a Conference in Honor of T. G. Ostrom
83. *G. I. Zapata,* Functional Analysis, Holomorphy, and Approximation Theory
84. *S. Greco and G. Valla,* Commutative Algebra: Proceedings of the Trento Conference
85. *A. V. Fiacco,* Mathematical Programming with Data Perturbations II
86. *J.-B. Hiriart-Urruty, W. Oettli, and J. Stoer,* Optimization: Theory and Algorithms
87. *A. Figa Talamanca and M. A. Picardello,* Harmonic Analysis on Free Groups
88. *M. Harada,* Factor Categories with Applications to Direct Decomposition of Modules
89. *V. I. Istrăţescu,* Strict Convexity and Complex Strict Convexity: Theory and Applications
90. *V. Lakshmikantham,* Trends in Theory and Practice of Nonlinear Differential Equations
91. *H. L. Manocha and J. B. Srivastava,* Algebra and Its Applications
92. *D. V. Chudnovsky and G. V. Chudnovsky,* Classical and Quantum Models and Arithmetic Problems
93. *J. W. Longley,* Least Squares Computations Using Orthogonalization Methods
94. *L. P. de Alcantara,* Mathematical Logic and Formal Systems
95. *C. E. Aull,* Rings of Continuous Functions
96. *R. Chuaqui,* Analysis, Geometry, and Probability
97. *L. Fuchs and L. Salce,* Modules Over Valuation Domains
98. *P. Fischer and W. R. Smith,* Chaos, Fractals, and Dynamics
99. *W. B. Powell and C. Tsinakis,* Ordered Algebraic Structures
100. *G. M. Rassias and T. M. Rassias,* Differential Geometry, Calculus of Variations, and Their Applications
101. *R.-E. Hoffmann and K. H. Hofmann,* Continuous Lattices and Their Applications
102. *J. H. Lightbourne, III, and S. M. Rankin, III,* Physical Mathematics and Nonlinear Partial Differential Equations
103. *C. A. Baker and L. M. Batten,* Finite Geometries
104. *J. W. Brewer, J. W. Bunce, and F. S. Van Vleck,* Linear Systems Over Commutative Rings

Other Volumes in Preparation

LINEAR SYSTEMS OVER COMMUTATIVE RINGS

JAMES W. BREWER
JOHN W. BUNCE
F. S. VAN VLECK
The University of Kansas
Lawrence, Kansas

MARCEL DEKKER, INC. New York and Basel

Library of Congress Cataloging-in-Publication Data

Brewer, James W., [date]
Linear systems over commutative rings.

(Lecture notes in pure and applied mathematics ; 104)
Bibliography: p.
Includes index.
1. Commutative rings. 2. Algebras, Linear.
I. Bunce, John W. [date]. II. Van Vleck, Fred S.
III. Title. IV. Series.
QA251.3.B72 1986 512'.4 86-4383
ISBN 0-8247-7559-7

MARCEL DEKKER, INC.
270 Madison Avenue, New York, New York 10016

Current printing (last digit):
10 9 8 7 6 5 4 3 2 1

PRINTED IN THE UNITED STATES OF AMERICA

Preface

Linear systems over commutative rings have been intensively studied for the past fifteen years. Much of this work was carried out by mathematicians and theoretical engineers studying linear systems depending on a parameter or linear systems with delays. The use of algebraic methods has clarified and unified the study of linear systems of these types.

In the last ten years three important surveys of the area have appeared. These are due to E. D. Sontag [59], E. W. Kamen [33], and G. Naudé and C. Nolte [45]. Each of these publications has "Linear Systems Over Rings" as part of its title. Each of these three papers has heavily influenced this book. In fact, our first exposure to most of the material in this book came from reading one or the other of these surveys. Another paper that heavily influenced the writing of this book was that of Bumby et al. [10]. In reading and filling in the details of these papers, we came to feel that there was a need to give a complete, self-contained treatment of this subject. In this book we attempt to bring together in one place the statements and proofs of the fundamental results of linear systems over rings. Our intended audience in this endeavor is all those interested in the mathematical structure of the subject, including graduate students in mathematics, mature mathematicians who are not specialists in the area, as well as mathematically inclined systems and control theorists.

We have striven for as much generality as possible, but have attempted to use only basic algebraic tools. Except for showing that certain rings which occur naturally in analysis have nice algebraic properties, we have in the main used only algebraic techniques. The decision to do so has restricted our selection of topics to be included.

Many of our arguments are easy extensions to module theory of classical linear algebra arguments. For the nonalgebraists we have included a chapter, Chapter 1, on commutative algebra; for others this chapter can be skipped.

We include a few exercises at the end of each chapter. Some of these are routine exercises that cover selected concepts of the chapter. Others outline proofs that are omitted in the text. A few indicate limitations of the theory.

We conclude each chapter with a section on "Notes and Remarks." In these notes we give references to the sources from which we learned the material in the chapter. We also try to give a little historical detail about the development of the theorems. Since engineers, control theorists, and algebraists have all contributed to the algebraic theory of linear systems, the literature is scattered throughout journals of various types. We will undoubtedly make attribution errors and we apologize for them in advance.

We are indebted to many people for their help, their comments, and their encouragement while we were writing this book. In particular, we are grateful to Paul Conrad, Philip Montgomery, Douglas Weakley, and Michael Darnel for their suggestions and criticisms when some of this material was presented in a seminar. William Heinzer, Daniel Katz, David

Lantz, and William Ullery helped us with various parts of Chapter 3. Irving Kaplansky provided us with the proofs of Theorems 3.14 and 3.15 and Robert Gilmer came to our assistance in showing that the ring of real analytic functions is infinite dimensional (see Theorem 3.17). Jimmy Arnold helped us to understand power series and rationality. The first author is grateful to Gert and Cornelia Naudé for several enlightening conversations as well as an initial critique of the manuscript. We are also grateful to Ed Kamen and Eduardo Sontag both for their correspondence and for providing us with several valuable preprints. We thank the unknown referees for their valuable comments on a preliminary version of this book.

We give special thanks to Carol Johnson for typing the preliminary version of this book and to Sharon Gumm for her careful preparation of the camera ready copy.

James W. Brewer
John W. Bunce
F. S. Van Vleck

Contents

LINEAR SYSTEMS OVER COMMUTATIVE RINGS

0
Introduction

This book is intended to demonstrate the use of commutative algebra in the theory of linear systems. Much of the mathematics that we discuss was developed by theoretical engineers. In this introduction we briefly indicate how the abstract structures that we consider arise in theoretical engineering. In the rest of the book we consider the structures abstractly, only occasionally making reference to problems that originally motivated the considerations. In Chapter 1 we develop the material from commutative algebra that we need. Readers who are experienced in commutative algebra should start in Chapter 2. We have tried to make it possible for other readers to read Chapters 2, 3, 4 with an occasional reference back to Chapter 1. For those who wish to systematically go through the entire book, algebraic maturity equivalent to a first-year graduate algebra course should suffice—in particular, we assume that the reader is acquainted with such notions as ring, ideal, and module.

The equation of motion of a simple pendulum, as in Figure 1, with gravity the only external force is

$$mr\,\frac{d^2x}{dt^2} + br\,\frac{dx}{dt} + mg\sin x(t) = 0$$

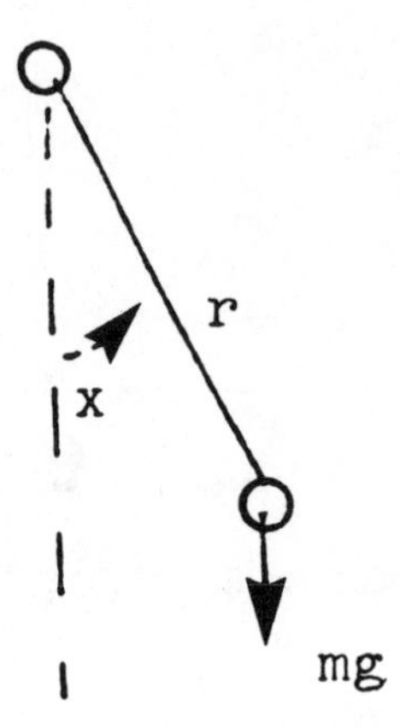

Figure 1

where m is the mass of the bob, r is the length of the massless rod, g is the acceleration of gravity, and b is a coefficient of friction (air resistance).

If we linearize the problem by replacing sin x by x and introduce new variables

$$\bar{x} = \begin{bmatrix} x_1 \\ x_2 \end{bmatrix}$$

with $x_1 = x$ and $x_2 = dx/dt$, then the second order equation is replaced by the following linear system:

$$\frac{d\bar{x}}{dt} = F\bar{x}(t)$$

where F is the 2 × 2 scalar matrix

$$\begin{bmatrix} 0 & 1 \\ -\frac{g}{r} & -\frac{b}{m} \end{bmatrix}$$

The state of the system is described by $\bar{x}$, which gives the angular displacement and angular velocity of the bob.

We now consider the more complicated system of Figure 2. In this case the pivot of the pendulum is attached to a collar that is around a rod rotating with angular velocity u(t). We assume that u(t) is something we can control; that

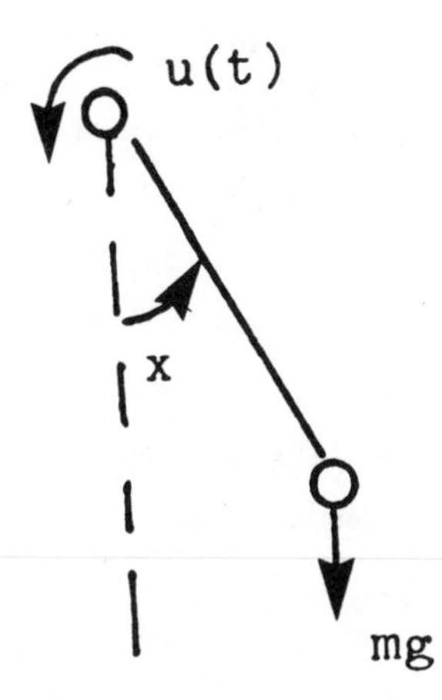

Figure 2

is, u(t) is an input. Assume that the coefficient of friction between the collar and the rod is k. The equation of motion is

$$mr\,\frac{d^2x}{dt^2} = -mg\sin x - br\,\frac{dx}{dt} + k\left(u - \frac{dx}{dt}\right)$$

$$\text{or}\qquad mr\,\frac{d^2x}{dt^2} + (br + k)\,\frac{dx}{dt} + mg\sin x = ku \tag{1}$$

If we again replace sin x by x and set

$$\bar{x} = \begin{bmatrix} x_1 \\ x_2 \end{bmatrix}, \quad x_1 = x, \quad x_2 = \frac{dx}{dt}$$

then this second order equation is replaced by the following linear system:

$$\frac{d\bar{x}}{dt} = F\bar{x}(t) + Gu(t) \tag{2}$$

where

$$F = \begin{bmatrix} 0 & 1 \\ -\frac{g}{r} & c \end{bmatrix}, \quad G = \begin{bmatrix} 0 \\ d \end{bmatrix}, \quad c = -(b + \frac{k}{r})/m, \text{ and } d = \frac{k}{mr}$$

The state of this system is described by $\bar{x}$, and the single input is u(t).

Suppose that the rotating rod is equipped with a mechanism that automatically rotates the rod at a rate proportional to dx/dt. That is, assume $u(t) = q\, dx/dt + w(t)$, where $w(t)$ is the angular velocity given to the rod by some other mechanism. Equation (1) linearized then becomes

$$mr \frac{d^2x}{dt^2} + (br + k - kq) \frac{dx}{dt} + mg\, x = kw \tag{3}$$

Clearly any type of damping can be achieved by the proper choice of q. The term $q\, dx/dt$ is called a feedback term. Letting K be the 1×2 matrix $[0 \quad -q]$, the first order linear system corresponding to (3) is

$$\frac{d\bar{x}}{dt} = (F - GK)\bar{x} + Gw$$

where F and G are the same matrices as in equation (2). The matrix K is called the *feedback matrix*. An important problem in control theory is that of constructing feedback matrices so that the perturbed matrix $F - GK$ has nice properties.

We now take the Laplace transform of equation (2), assuming that $\bar{x}(0) = 0$. Let $X_i(s)$ be the Laplace transform of $x_i(t)$ and let $U(s)$ be the Laplace transform of $u(t)$. (We assume that all the functions involved behave nicely under Laplace transforms.) We get the transformed equation

$$sX(s) = FX(s) + GU(s) \tag{4}$$

where

$$X = \begin{bmatrix} X_1 \\ X_2 \end{bmatrix}$$

Solving (4) we see that formally

$$X(s) = (sI - F)^{-1}GU(s)$$

If the absolute value of s is greater than the operator norm of the matrix F, then the inverse of $(sI - F)$ actually exists, and expanding by the geometric series we see that

$$X(s) = \left(\sum_{n=1}^{\infty} F^{n-1}Gs^{-n} \right) U(s)$$

We set

$$T(s) = \sum_{n=1}^{\infty} F^{n-1}Gs^{-n}$$

and call T(s) the *transfer function* of the system (2). In engineering applications the transfer function is often the object of study. Since the matrix sequence $\{F^{n-1}G\}_1^{\infty}$ determines the transfer function, it is fundamental to our algebraic treatment. We shall study it extensively.

In the above considerations, the matrices involved all had real entries. As we shall now see, one realistic model that occurs in applications requires matrices with entries in the ring of polynomials in one indeterminate over a field. In equation (3) above we assumed that the feedback term was

$$q \frac{dx}{dt}(t)$$

that is, we assumed that the mechanism responded instantaneously. In practice the control often requires time $a > 0$ to respond, and the feedback term is

$$q \frac{dx}{dt}(t - a)$$

Equation (3) then becomes

$$mr \frac{d^2x}{dt^2}(t) + (br + k) \frac{dx}{dt}(t) - kq \frac{dx}{dt}(t - a) + mg\, x(t) = kw(t) \qquad (5)$$

An equation of this type is called a delay differential equation. A great deal of research has been done on delay differential equations; see [5] or [6]. Until recently, most of this research used analytic or functional analytic methods. The discovery that some problems involving systems with time delays can be studied using algebraic techniques is due to Kamen [32]. The first-order linear system corresponding to (5) is

$$\frac{d\bar{x}}{dt} = F\bar{x}(t) - GK\bar{x}(t - a) + Gw(t) \tag{6}$$

This system is not really a finite-dimensional system. The solution to (6) is not determined uniquely by the usual initial conditions $\bar{x}(0) = \bar{x}_0$. Rather, $\bar{x}$ must be specified on the entire interval $-a \leq t \leq 0$; see [16].

We now take the Laplace transform of equation (6). Assuming that $x(\bar{t}) = 0$ for $-a \leq t \leq 0$, we get that

$$sX(s) = FX(s) - e^{-as}\, GKX(s) + GW(s)$$

$$X(s) = (sI - (F - e^{-as}\, GK))^{-1}\, GW(s)$$

The transfer function in this case is

$$T(s) = \sum_{n=1}^{\infty} (F - e^{-as}\, GK)^{n-1}\, Gs^{-n}$$

The matrix sequence $\{(F - e^{-as}\, GK)^{n-1}G\}_1^{\infty}$ that determines the transfer function now has entries in the ring of polynomials in the indeterminate e^{-as}. We remark that if s is an indeterminate over **R** then so is e^{-as}; that is, e^{-as} does not satisfy a polynomial over **R**. The classical theory of linear systems over fields does not apply in this case. We use techniques from commutative ring theory to study abstract linear systems over rings in Chapters 2, 3, and 4.

In general a system might involve many components with time delays. The delays are called commensurate if all of them are integral multiples of some number. In this case the equation of the system is

$$\frac{dx}{dt} = \sum_{i=0}^{q} F_i x(t - ia) + \sum_{i=0}^{r} G_i u(t - ia) \qquad (7)$$

In this equation a is a positive number, the F_i are $n \times n$ real matrices, the G_i are $n \times m$ real matrices, and $x(t)$, $u(t)$ are column vectors over the reals **R**. The system (7) can be viewed directly as a system over **R**[d], d an indeterminate, as follows: Let

$$V = \{f: \mathbf{R} \longrightarrow \mathbf{R} \mid \text{there is a } t_f \text{ such that } f(t) = 0 \text{ for all } t < t_f\}$$

Define $d : V \to V$ by $(df)(t) = f(t - a)$, and use the same symbol to denote the corresponding operator on column vectors with entries from V. With this notation equation (7) becomes

$$\frac{dx}{dt} = \sum_{i=1}^{q} F_i (d^i x)(t) + \sum_{i=1}^{r} G_i (d^i u)(t)$$

Let

$$F(d) = \sum_{i=0}^{q} F_i d^i, \qquad G(d) = \sum_{i=0}^{r} G_i\, d^i$$

The equation then becomes

$$\frac{dx}{dt} = (F(d)x)(t) + (G(d)u)(t)$$

where F(d) and G(d) are matrices over the ring **R**[d]. If not all of the delays were commensurate, then the matrices corresponding to F(d) and G(d) would have entries in a ring of polynomials in more than one indeterminate.

Linear systems theory over rings can also be applied to systems depending on parameters. Such systems might arise in several ways. For instance, the constant coefficients in a differential equation might be functions of one or several parameters. The matrices describing such an equation would have entries in a ring of functions. If the entries depend analytically on a real parameter, then the ring would be the ring of analytic functions on the real numbers **R**. Another place where such systems arise is in the study of nonlinear systems. Linearizing the nonlinear system at different points would lead to coefficients that are functions of the points of linearization, the parameters in this case. There have been several recent studies of systems depending on parameters; see, for example, [11] and [61].

In applications one cannot always observe the system itself; instead one observes or measures certain particular aspects of the system. The quantities measured are called outputs $y(t)$, and we assume that the vector $y(t)$ is obtained from the state $x(t)$ in a linear fashion, $y(t) = Hx(t)$. Of course, it may be the case that $y(t)$ depends on $x(t)$ at some earlier time. In this case the equations for the system are

$$\begin{aligned} \frac{dx}{dt} &= Fx(t) + Gu(t) \\ y(t) &= Hx(t) \end{aligned} \tag{8}$$

where $x(t)$, $y(t)$ and $u(t)$ are column vectors, and F, G, and H are $n \times n$, $n \times m$, and $p \times n$ (respectively) matrices with entries in some polynomial ring over **R**. This is an m-*input*, p-*output*, n-*dimensional*, *continuous-time*, *time-invariant linear system*. Time invariance means that the entries of F,G and H are independent of t.

Intuitively, the system described by (8) is called *observable* if one can determine the state of the system by looking at the outputs. We will study observability and its dual concept, reachability, in Chapter 2.

We now go back to equation (2) and add an output equation:

$$\frac{dx}{dt} = Fx(t) + Gu(t)$$

$$y(t) = Hx(t)$$

Here we assume that the matrices H,F, and G have real entries. Instead of writing down the differential equation for the system, one often speaks of the system (H,F,G). If we assume that $x(0) = 0$ and take the Laplace transform of this system as before, we get

$$Y(s) = \left(\sum_{i=1}^{\infty} HF^{i-1}Gs^{-i} \right) U(s)$$

The matrix sequence $\{A_i\}_{i=1}^{\infty}$, where $A_i = HF^{i-1}G$, thus determines the transfer function, and hence the output y that results from each input u. (Of course, one goes from Y(s) back to y(t) by taking the inverse Laplace transform.) The sequence $\{A_i\}$ is called an *input-output map*, or more concisely, an i/o *map*. A natural question is the following: Given $\{A_i\}$ (that is, given a transfer function), when does there exist a linear system (H,F,G) such that $A_i = HF^{i-1}G$ for each i? That is, when is an abstractly given transfer function actually realized by some linear system? This is called the *realization problem*. We study this problem in Chapter 4.

If $dx/dt = Fx(t)$ is a system of first order differential equations with F an $n \times n$ real matrix and if F has distinct eigenvalues $r_1, r_2, \ldots, r_n$ with associated eigenvectors $v_1, v_2, \ldots, v_n$, then the general (possibly complex-valued) solution of the system is

$$x(t) = \sum_{i=1}^{n} C_i e^{r_i t} v_i$$

where the C_i are arbitrary real or complex numbers. If the r_i all have negative real parts, then

$$\lim_{t\to\infty} x(t) = 0$$

so that every solution of the homogeneous (unforced) system $dx/dt = Fx(t)$ goes to zero as t approaches infinity. Since every solution of the (forced) nonhomogeneous system $dx/dt = Fx(t) + w(t)$, $x(t_0) = x_0$, can be written as

$$x(t) = x_h(t) + x_p(t)$$

where $x_h(t)$ is the solution of $dx/dt = Fx(t)$ satisfying $x_h(t_0) = x_0$ and $x_p(t)$ is the solution of $dx/dt = Fx(t) + w(t)$ satisfying $x_p(t_0) = 0$, we see that for $Re(r_i) < 0$, the influence of the initial condition x_0 is negligible for large values of t. It is hence desirable for all eigenvalues to be in the open left half-plane. In the case of a system like $dx/dt = Fx(t) + Gu(t)$ with control term $Gu(t)$, one would like to be able to choose a feedback matrix K so that, letting $u(t) = -Kx(t) + w(t)$, the system

$$\frac{dx}{dt} = (F - GK)x(t) + Gw(t)$$

has the eigenvalues of $(F - GK)$ in the open left half-plane. The problem of finding a matrix K so that $F - GK$ has all of its eigenvalues in the open left half-plane is called the *stabilization problem*. The problem of finding a matrix K so that the eigenvalues of $F - GK$ are exactly specified is called the *pole assignability problem*. We study these problems in Chapter 3.

Finally, we note that the structures we will consider also occur in time-invariant, discrete-time systems given by equations

$$x(k + 1) = Fx(k) + Gu(k)$$
$$y(k) = Hx(k) \tag{9}$$

where k is an integer and F,G and H are $n \times n$, $n \times m$, and $p \times n$ (respectively) matrices over a commutative ring. Systems of this type arise in coding theory and automata theory. If $x(0) = 0$, then iteration of (9) shows that for $n > 0$,

$$x(n) = \sum_{i=0}^{n-1} F^i Gu(n - 1 - i)$$
$$y(n) = \sum_{i=0}^{n-1} HF^i Gu(n - 1 - i) \tag{10}$$

Thus, the i/o sequence $HF^{i-1}G$, $i \geq 1$, again determines the output from given input. Also, equations (10) show that any state x(n) can be reached, starting from $x(0) = 0$, by the proper choice of inputs if and only if the map determined by the $n \times (nm)$ matrix $[G, FG, \ldots, F^{n-1}G]$ is onto. This property is called *reachability* and, as noted before, is treated in Chapter 2.

EXERCISES

1. Consider the delay differential equation

$$x(t) = x_0, \quad 0 \leq t \leq a$$

$$\frac{dx}{dt} = -cu_a(t)x(t - a), \quad 0 \leq t$$

where

$$u_a(t) = \begin{cases} 0 & \text{for} \quad t < a \\ 1 & \text{for} \quad a \leq t \end{cases}$$

Show that X(s), the Laplace transform of x(t), is

$$X(s) = \frac{x_0}{s + ce^{-as}}$$

Show that, for large s,

$$X(s) = x_0 \sum_{n=1}^{\infty} (-1)^{n-1} c^{n-1} (e^{-as})^{n-1} s^{-n}$$

so that the coefficients of s^{-n} are polynomials in the indeterminate e^{-as}. (This differential equation models the mixing of liquids in a tank when the mixing is not instantaneous; see [16, chapter V].)

2. Prove that if b is a nonzero constant and s is an indeterminate over **R**, then so is e^{-bs}; that is, e^{-bs} does not satisfy a polynomial over **R**. (Hint: If $a_n e^{-nbs} + \cdots + a_1 e^{-bs} + a_0 = 0$ for all s, take limits as $s \to -\infty$ (if $b > 0$) or as $s \to -\infty$ (if $b < 0$), and conclude that $a_0 = 0$.)

3. Consider the nth order constant coefficient linear differential equation

$$x^{(n)}(t) + \sum_{k=0}^{n-1} a_k x^{(k)}(t) = u(t) \qquad (*)$$

where $x^{(k)}(t)$ devotes the kth derivative of x(t). Let $\bar{x}(t)$ devote the column vector with entries x(t), x′(t), ..., $x^{(n-1)}(t)$. Consider u(t) as input and x(t) as output. Find the matrices H, F, and G that give (*) as a first-order linear system. What is the characteristic polynomial of F?

4. Let x′′ + ax′ + bx = 0 be a constant-coefficient differential equation. Prove that every solution goes to zero as

$t \longrightarrow \infty$ if and only if both a and b are greater than zero.

5. Consider the transfer function

$$T(s) = \frac{1}{(s-1)(s+e^{-s})}\begin{bmatrix} s-1 & e^{-s}s+1+e^{-2s} \\ 1-s & -1 \end{bmatrix}$$

Show that the first few terms of the i/o map associated with T(s) are

$$A_1 = \begin{bmatrix} 1 & e^{-s} \\ -1 & 0 \end{bmatrix}$$

$$A_2 = \begin{bmatrix} -e^{-s} & 1+e^{-s} \\ e^{-s} & -1 \end{bmatrix}$$

$$A_3 = \begin{bmatrix} e^{-2s} & 1 \\ -e^{-2s} & e^{-s}-1 \end{bmatrix}$$

(Recall that A_i is the coefficient of s^{-i} in the expansion of T(s).)

6. Prove that the map determined by the matrix $[G, FG, \ldots, F^{n-1}G]$ is onto if and only if any x(n) in equation (10) can be reached, starting from x(0) = 0, by the proper choice of imputs u(0),u(1),...,u(n - 1).

1
Algebraic Preliminaries

In this chapter we recall some of the areas of commutative algebra that we shall need in the sequel. A commutative algebraist can omit this chapter altogether, while the non-algebraist might wish to refer to it as needed. On the other hand, someone desiring a complete treatment would be advised to read the chapter before continuing.

We have not tried to be encyclopedic, but have tried to touch most of the important areas and to prove most of the facts we shall use. There are several excellent references on commutative algebra and we have drawn freely from them. The interested reader can consult the References where several texts are listed.

We assume that the reader is familiar with the notions of ring, ideal, module, etc., and that he is acquainted with the notions of free module, polynomial ring, prime ideal, maximal ideal, and exact sequence, to name a few. For us, all rings will be commutative with identity and all modules unitary.

1.1 THE CAYLEY-HAMILTON THEOREM

The Cayley-Hamilton Theorem is one of the most useful results of classical linear algebra over fields. It is also valid and useful over an arbitrary commutative ring.

Let R be a commutative ring. Then much of the ordinary theory of determinants carries over to matrices over R. For example, the classical adjoint formula holds:

If A is an $n \times n$ matrix over R, then $A \cdot \operatorname{adj} A = \det(A)I$, where adj A denotes the adjoint of A and det(A) denotes the determinant of A.

Also, if A is an $n \times n$ matrix over R, and if X is an indeterminate, we can define the *characteristic polynomial* of A to be $\det(XI - A)$, where I is the $n \times n$ identity matrix. Clearly, $\det(XI - A)$ is a monic polynomial in the polynomial ring R[X], say

$$\det(XI - A) = X^n + r_{n-1}X^{n-1} + \cdots + r_1X + r_0$$

The Cayley-Hamilton Theorem asserts that the matrix

$$A^n + r_{n-1}A^{n-1} + \cdots + r_1A + r_0$$

is the zero matrix—that is, A satisfies its characteristic polynomial. The intent of this section is to prove this beautiful result. Our method involves some elegant algebra and also a very brief departure from our convention that all rings are commutative.

Let R be a commutative ring with X an indeterminate and n a positive integer. The set R_n of all $n \times n$ matrices over R is a (noncommutative) ring under ordinary addition and multiplication of matrices. Moreover, there is a natural isomorphism between $R_n[X]$ = the ring of all polynomials over the ring R_n in the commuting indeterminate X and $(R[X])_n$ = the ring of all $n \times n$ matrices over the ring R[X]. We shall exploit this fact in the proof of the following theorem.

THEOREM 1.1. (Cayley-Hamilton). Let R be a commutative ring with A an $n \times n$ matrix over R. Then A satisfies its characteristic polynomial.

Proof. Let X be an indeterminate. By the classical

adjoint formula applied to the matrix $XI - A \in (R[X])_n$, we have that

$$(XI - A)\,\mathrm{adj}(XI - A) = \det(XI - A)I = f(X)I \qquad (*)$$

where $f(X)$ is the characteristic polynomial of A. Note that not only does $f(X) \in R[X]$, but also to $R_n[X]$ once we embed R into R_n by the injection $a \rightarrow aI$. Therefore, via the isomorphism between $(R[X])_n$ and $R_n[X]$, equation (*) says precisely that we have a factorization of $f(X)$ in the ring $R_n[X]$.

We will be able to complete the proof once we have a strong version of the "Factor Theorem": Let S be a (not necessarily commutative) ring with $s \in S$. If $g(Y)$ is a polynomial in $S[Y]$, then s is a root of g if and only if $Y - s$ is a factor of g. To see this, observe that due to the simple form of $Y - s$, we can carry out the steps of the Division Algorithm to obtain the equation

$$g(Y) = (Y - s)q(Y) + r \quad \text{for } r \in S$$

If s is a root of g, then $0 = g(s) = (s - s)q(s) + r$. Hence, $Y - s$ is a factor of $g(Y)$. The converse is obvious.

In our situation, we have that $XI - A$ is a factor of $f(X)$ in the ring $R_n[X]$. (Recall that I is the identity element of the ring R_n.) It follows from the Factor Theorem that A is a root of $f(X)$ as desired.

The observant reader will notice that we have only used the "obvious" part of the Factor Theorem in the proof of Theorem 1.1.

1.2 NOETHERIAN RINGS AND NOETHERIAN MODULES

A ring R is said to be *noetherian* if each ideal of R is finitely generated. This is easily seen to be equivalent to the condition: If $I_0 \subseteq I_1 \subseteq \ldots$ is an ascending chain of ideals of R, then there exists a positive integer n such that $I_j = I_{n+1}$ for $j \geq n$. In some quarters, noetherian rings are regarded as being the most important class of rings. We shall content ourselves with proving a few of their properties.

Our first theorem, which says that the noetherian property is preserved under polynomial ring formulation, is due originally to Hilbert.

THEOREM 1.2. Let R be a ring with $X_1, \ldots, X_n$ indeterminates. If R is noetherian, so is $R[X_1, \ldots, X_n]$.

Proof. We can by induction assume that $n = 1$. Thus, let J be an ideal in R[X] and let I_n be the set of leading coefficients of polynomials of degree $\leq n$ in J together with 0. Then I_n is an ideal of R and

$$J \cap R = I_0 \subseteq I_1 \subseteq I_2 \subseteq \cdots$$

Set $I = \cup I_n$. Let $f_1, \ldots, f_k$ be polynomials in J whose leading coefficients generate I. Suppose that N is the maximum of the degrees of the f's. For $0 \leq j \leq N - 1$, pick a finite number of polynomials $g_{j1}, g_{j2}, \ldots$ in J whose leading coefficients generate I_j. A calculation shows that the f's together with the g's generate J.

As a corollary to Theorem 1.2 we see that the rings $Z[X_1, \ldots, X_n]$ and $L[X_1, \ldots, X_n]$ are noetherian, where Z is the ring of integers and L is any field.

There is another fact about noetherian rings and modules over them that we need to prove. To do so, we have to introduce the notion of a noetherian module. If R is a ring with M a module over R, then we say that M is a *noetherian module* if each submodule of M is finitely generated. This is equivalent to the *ascending chain condition*: If $M_0 \subseteq M_1 \subseteq \cdots$ is a chain of submodules of M, then there exists a positive integer

N such that $M_j = M_{j+1}$ for $j \geq N$. Another equivalent formulation is: An R-module M is noetherian if and only if it satisfies the *maximum condition*—that is, each nonempty family of submodules of M contains a maximal element.

Before proving our main result on noetherian modules, we require two lemmas.

LEMMA 1.3. Let R be a commutative ring with M an R-module. If A and B are submodules of M such that A/B and (A + B)/A are finitely generated, then B is finitely generated.

Proof. For $m \in M$, let $\bar{m}$ denote the coset m + A. Let $\bar{b}_1,\ldots,\bar{b}_n$ generate (A + B)/A and $b_{n+1},\ldots,b_t$ generate $A \cap B$. If $b \in B$, then $\bar{b} = \sum_{i=1}^{n} r_i\bar{b}_i$ for some elements $r_i \in R$. Hence

$$b - \sum_{i=1}^{n} r_i b_i \in A \cap B$$

Therefore,

$$b - \sum_{i=1}^{n} r_i b_i = \sum_{i=n+1}^{t} s_i b_i$$

and it follows that

$$b = \sum_{i=1}^{n} r_i b_i + \sum_{i=n+1}^{t} s_i b_i$$

Thus, $b_1,\ldots,b_t$ generate B.

LEMMA 1.4. Let R be a commutative ring with

$$0 \longrightarrow M' \xrightarrow{\alpha} M \xrightarrow{\beta} M'' \longrightarrow 0$$

an exact sequence of R-modules. Then M is noetherian if and only if M′ and M′′ are noetherian.

Proof. Suppose that M is noetherian. If N′ is a submodule of M′, then $\alpha(N')$ is a submodule of M and hence is finitely generated. Since α is injective, $\alpha(N') \cong N$ and so N' is finitely generated. If N'' is a submodule of M'', then $\beta^{-1}(N'')$ is a submodule of M and hence is finitely generated. Since β is surjective, $N'' = \beta(\beta^{-1}(N''))$ and consequently is finitely generated.

Conversely, suppose that M' and M'' are noetherian and that N is a submodule of M. Then $\alpha^{-1}(N)$ is a submodule of M' and $\beta(N)$ is a submodule of M''. Letting α^* and β^* be the restrictions, this induces the exact sequence

$$0 \longrightarrow \alpha^{-1}(N) \xrightarrow{\alpha^*} N \xrightarrow{\beta^*} \beta(N) \longrightarrow 0$$

with $\alpha^{-1}(N)$ and $\beta(N)$ finitely generated since M' and M'' are noetherian. This is the set-up of Lemma 1.3 with $A = \alpha^*\alpha^{-1}(N)$ and $B = N$.

We are now in a position to prove the following fundamental result.

THEOREM 1.5. A finitely generated module over a noetherian ring is a noetherian module.

Proof. Let R be a noetherian ring with M a finitely generated R-module, say M can be generated by n elements. Then there exists an exact sequence of the form

$$0 \longrightarrow M' \longrightarrow R^n \longrightarrow M \longrightarrow 0$$

By Lemma 1.4, to see that M is noetherian, we have only to see that R^n is a noetherian R-module. For $n = 1$, this is the definition of noetherian ring. By way of induction, suppose that R^{n-1} is a noetherian R-module and consider the exact sequence

$$0 \longrightarrow R \longrightarrow R^n \longrightarrow R^{n-1} \longrightarrow 0$$

where the maps are the obvious ones. That R^n is noetherian follows also from Lemma 1.4. This completes the proof of of Theorem 1.5.

We immediately put Theorem 1.5 to use to help us prove one of the most important properties that noetherian rings have. It is the following: If an ideal I of a noetherian ring R consists entirely of zero divisors, then there exists a single nonzero element $r \in R$ such that $rx = 0$ for each element $x \in I$. We begin by proving a useful combinatorial fact about prime ideals.

THEOREM 1.6. Let R be a commutative ring with $P_1, P_2, \ldots, P_n$ prime ideals of R. If A is an ideal of R and if $A \subseteq \bigcup_{i=1}^{n} P_i$, then $A \subseteq P_j$ for some j.

Proof. It is harmless to assume that there are no containment relations among the P's. Suppose that A is contained in no P. If a prime ideal contains a product (or intersection) of ideals, it contains one of them. Consequently, for $1 \leqslant i \leqslant n$, there exists an element

$$a_i \in [A \cap (\bigcap_{j \neq i} P_j)] \backslash P_i$$

Consider the element $a = \sum_{i=1}^{n} a_i$. Then $a \in A$ and if, for example, $a \in P_1$, then $a - (\sum_{i=2}^{n} a_i) = a_1 \in P_1$, a contradiction. This argument applies equally to each of the P's and so a is an element of A belonging to no P. This contradiction proves the theorem.

Since it is not more difficult to prove, we give a stronger result than we need.

THEOREM 1.7. Let R be a noetherian ring with M a finitely generated R-module. Then the set of elements of R which are zero divisors on M is a finite union of prime ideals of R. Moreover, if $P_1, \ldots, P_n$ is the set of prime ideals involved, then each P_i is the annihilator of a single nonzero element of M.

Proof. Consider the set of annihilators of nonzero elements of M. Each annihilator is contained in a maximal one by the ascending chain condition on R. Evidently, the set of all zero divisors on M is the set-theoretic union of these maximal ones. We first claim that each of the maximal annihilators is a prime ideal. To see this, let I be a maximal annihilator and suppose that I is the annihilator of $m \in M$, $m \neq 0$. Given $ab \in I$, we must prove that a or b lies in I. Assume that $a \notin I$. Then $am \neq 0$. Notice also that the annihilator J of am contains I. Since I is maximal among annihilators of single elements of M, $J = I$. Now $b \cdot am = abm = 0$ and so $b \in J = I$. Therefore, the claim is justified.

We next claim that there are only finitely many such maximal annihilators of single elements. Denote them by $\{P_i\}$ and let P_i be the annihilator of m_i. Since M is a noetherian module, the submodule of M generated by the m's can be generated by finitely many of the m's, say $m_1, \ldots, m_t$. If any further m's exist we have an equation

$$m_{t+1} = r_1 m_1 + \cdots + r_t m_t \quad \text{for} \quad r_i \in R$$

From this it follows that $P_1 \cap \cdots \cap P_t \subseteq P_{t+1}$. But $P_1 \cap \cdots \cap P_t \supseteq P_1 \cdots P_t$, and if a prime ideal contains a product of ideals, it contains one of them. Thus, some $P_j \subseteq P_{t+1}$, contradicting the maximality of P_j. Hence, there are no further m's or P's.

We are now in a position to prove the result we were after.

THEOREM 1.8. Let R be a noetherian ring with I an ideal of R consisting entirely of zero divisors of R. Then there exists a nonzero element $r \in R$ such that $rx = 0$ for each element $x \in I$.

Proof. In the notation of Theorem 1.7, suppose that M is the ring R itself. Then in the notation of Theorem 1.7, $I \subseteq \bigcup_{j=1}^{t} P_j$, and by Theorem 1.6, $I \subseteq P_j$ for some j. But P_j is the annihilator of the element $m_j \in R$ and so $m_j x = 0$ for for each element $x \in I$. This completes the proof.

1.3 INTEGRAL DEPENDENCE AND INTEGRAL CLOSURE

In this section we discuss the notions of integral dependence and, its weakened form, almost integral dependence. These ideas are important in commutative algebra and we shall use them often in the sequel. The basic set-up is as follows. We have a pair of rings $R \subseteq S$ with R a subring of S. Among other things, this means that the rings share the same identity element. An element $s \in S$ is said to be *integral over* R if there is a monic relation of the form

$$s^n + r_{n-1}s^{n-1} + \cdots + r_1 s + r_0 = 0$$

for some positive integer n and elements $r_0, \ldots, r_{n-1} \in R$. We immediately have an alternate useful characterization. Note that the proof again uses the theory of determinants over a commutative ring.

THEOREM 1.9. Let $R \subseteq S$ be rings with $s \in S$. The following conditions are equivalent.

1. The element s is integral over R.
2. The ring R[s] is a finitely generated R-module.
3. The element s belongs to a subring T of S that contains R and is a finitely generated R-module.

Proof. (1) $\Longrightarrow$ (2): Suppose that s satisfies the relation

$$s^n + r_{n-1}s^{n-1} + \cdots + r_1 s + r_0 = 0$$

We claim that the ring $R[s]$ is generated as an R-module by the elements $1,\ldots,s^{n-1}$. To prove this, we have only to verify that, for $k \geq 0$,

$$s^k \in R \cdot 1 + R \cdot s + \cdots + R \cdot s^{n-1} = T$$

This is clearly the case for $0 \leq k \leq n - 1$ and, in fact,

$$s^n = (-r_0) + (-r_1)s + \ldots + (-r_{n-1})s^{n-1}$$

But then

$$\begin{aligned} s^{n+1} &= (-r_0)s + (-r_1)s^2 + \cdots + (-r_{n-1})s^n \\ &= (-r_0)s + (-r_1)s^2 + \cdots + (-r_{n-1})(-r_0 + \cdots \\ &\qquad + (-r_{n-1})s^{n-1}) \in T \end{aligned}$$

Obviously, an induction argument will finish the proof.

(2) $\Longrightarrow$ (3): Take $T = R[s]$.

(3) $\Longrightarrow$ (1): Let T be generated as an r-module by the elements $t_1,\ldots,t_n$. Then for $1 \leq i \leq n$, there exist elements $r_{i1},\ldots,r_{in} \in R$ so that $st_i = \sum_{j=1}^n r_{ij}t_j$. This gives rise to the system of equations

$$\begin{aligned} (s - r_{11})t_1 - r_{12}t_2 - \cdots - r_{1n}t_n &= 0 \\ -r_{21}t_1 + (s - r_{22})t_2 - \cdots - r_{2n}t_n &= 0 \\ -r_{n1}t_1 - r_{n2}t_2 - \cdots +(s - r_{nn})t_n &= 0 \end{aligned}$$

By Cramer's Rule, or perhaps more appropriately by its proof, if $d = \det[s\delta_{ij} - (r_{ij})]$ with δ_{ij} the Kronecker delta, then $dt_i = 0$ for $1 \leq i \leq n$. Thus, the element $d \in T$ annihilates each generator of the R-module T and hence annihilates T. But T is a subring of S and therefore contains the identity element 1 of S. It follows that

$$0 = d \cdot 1 = d = \det[s\delta_{ij} - (r_{ij})]$$

which is a monic polynomial in s with coefficients in R. Therefore, s is integral over R. This proves Theorem 1.9.

For later reference, we record a special instance of integral dependence.

THEOREM 1.10. Let R be a ring with X an indeterminate over R and suppose that $f(X) = X^n + r_{n-1}X^{n-1} + \cdots + r_1X + r_0$ is a monic polynomial over R. Let $S = R[X]/(f)$ and denote by x the coset $X + (f)$. Then x is integral over R and $\{1,x,\ldots,x^{n-1}\}$ is a free module basis for the ring $S = R[x]$ over R.

Proof. That x is integral over R follows from the fact that it satisfies the polynomial f(X). Moreover, as we saw in the first part of the proof of Theorem 1.9, the elements $1,x,\ldots,x^{n-1}$ generate the ring R[x] over R. Finally if $a_0 + a_1x + \cdots + a_{n-1}x^{n-1} = 0$ for elements $a_0,a_1,\ldots,a_{n-1} \in R$, then $a_0 + a_1X + \cdots + a_{n-1}X^{n-1}$ is divisible by f(X). This is clearly a contradiction unless each $a_i = 0$. Hence, $\{1,x,\ldots,x^{n-1}\}$ is a free basis.

The following corollary to Theorem 1.9 will also be of use to us.

THEOREM 1.11. Let $R \subseteq S$ be rings with a and b elements of S. If a and b are integral over R, then so are the elements $a - b$ and $a \cdot b$. In particular, the set of all elements of S integral over R is a subring of S.

Proof. By iterating the first part of the proof of Theorem 1.9, we see that the ring $(R[a])[b] = R[a,b]$ is a finitely generated module over the ring R[a]. Since R[a] is a finitely generated module over R, it follows that R[a,b] is a finitely generated R-module. Both $a - b$ and $a \cdot b$ belong to the ring R[a,b] and, by Theorem 1.9, are integral over R.

Let $R \subseteq S$ be rings. As proved in Theorem 1.11, the set of all elements of S integral over R forms a subring of S that clearly contains R. This subring is called the *integral closure of* R *in* S. If S is the integral closure of R in S, then we say that S *is integral over* R. We record the following corollary to Theorem 1.9.

THEOREM 1.12. Let $R \subseteq S \subseteq T$ be rings. If S is integral over R and if T is integral over S, then T is integral over R.

Proof. Let $t \in T$. Then there exist elements $s_0, \ldots, s_{n-1} \in S$ such that $t_n + s_{n-1}t^{n-1} + \cdots + s_0 = 0$. Since S is integral over R, each of $s_0, \ldots, s_{n-1}$ is integral over R. Iterated use of Theorem 1.9 shows that the ring $R[s_0, \ldots, s_{n-1}]$ is a finitely generated R-module. But then t belongs to the ring $R[s_0, \ldots, s_{n-1}, t]$, which is itself a finitely generated R-module. Another application of Theorem 1.9 shows that t is integral over R and completes the proof.

Let D be an integral domain—that is, a commutative ring having no zero divisors except 0. If L is the quotient field of D, we say that D is *integrally closed* if the integral closure of D in L is D. In other words, D is integrally closed if and only if each element of L integral over D belongs to D. For example, as we shall soon see, the ring Z of integers and the ring $L[X_1, \ldots, X_n]$ of polynomials over a field are both integrally closed. In fact, the rings Z and $L[X_1, \ldots, X_n]$ have a property that is stronger than integral closure. To make this statement precise, we must introduce a notion similar to that of integral dependence.

Again, let $R \subseteq S$ be rings. An element $s \in S$ is said to be *almost integral over* R if there exists a finitely generated R-submodule E of S such that $R[s] \subseteq E$. Evidently, $R[s] \subseteq E$ if and only if each nonnegative power of s belongs to E. If each element of S is almost integral over R, then we say that S is *almost integral over* R. Let D be an integral domain with quotient field L. If each element of L almost integral over D belongs to D, then we say that D is *completely integrally closed.* The terminology may be inconsistent, but it is standard.

Any integral element of S is almost integral over R. However, since the definition does not require that the R-submodule E be a subring of S, the notions of integral dependence and almost integral dependence need not coincide. Of course, if the ring R is noetherian, then by Theorem 1.5, the finitely generated R-module E is a noetherian module. Therefore, its submodule R[s] is a fintely generated R-module and s is integral over R. It follows that if R is noetherian and if s is almost integral over R, then s is integral over R. Hence, almost integral dependence and integral dependence coincide in case the ring R is noetherian.

We now expose some of the important properties of almost integral dependence.

THEOREM 1.13. Let $R \subseteq S$ be rings. If the elements $s_1,\ldots,s_n \in S$ are almost integral over R, then there exists a finitely generated R-submodule E of S such that $R[s_1,\ldots,s_n] \subseteq E$. Therefore, the set of all elements of S almost integral over R forms a subring of S, which is called the *complete integral closure of* R *in* S.

Proof. We argue by induction on n. If n = 1, the assertion is merely the definition. Suppose that the result is true for (n - 1) elements. Then there exists a finitely generated submodule $E_1 = Re_1 + Re_2 + \cdots + Re_\ell$ of S such that $R[s_1,\ldots,s_{n-1}] \subseteq E_1$. Also, since s_n is almost integral over R, there exists a finitely generated R-submodule $E_2 = Rf_1 + Rf_2 + \cdots + Rf_m$ of S such that $R[s_n] \subseteq E_2$. Since S is a ring, we can multiply the elements e_i and f_j, and it is easy to verify that $R[s_1,\ldots,s_n]$ is a submodule of the R-submodule of S generated by the set $\{e_if_j\}$, $1 \leqslant i \leqslant \ell$, $1 \leqslant j \leqslant m$. This completes the proof.

THEOREM 1.14. Let $R \subseteq S \subseteq T$ be rings. If S is almost integral over R and if T is integral over S, then T is almost integral over R.

Proof. Let $t \in T$. Since t is integral over S, there exist elements $s_0,\ldots,s_{n-1} \in S$ such that

$$t^n + s_{n-1}t^{n-1} + \cdots + s_1t + s_0 = 0$$

Thus, t is integral over the ring $R[s_0,\ldots,s_{n-1}] = R_0$. In fact, $\{1,\ldots,t^{n-1}]$ is an R_0-module generating set for $R_0[t]$ over R_0. By Theorem 1.13, there exists a finitely generated R-submodule $E = Re_1 + \cdots + Re_m$ of T such that $R_0 \subseteq E$. Then

$$R[t] \subseteq R_0[t] = R_0 + R_0t + \cdots + R_0t^{n-1} \subseteq \sum_{j=0}^{n-1} \sum_{i=1}^{m} Rt^je_i \subseteq T$$

Therefore, R[t] is contained in the R-submodule of T generated by $\{t^je_i\}$. It follows that t is almost integral over R.

Let D be an integral domain with quotient field L and let $x \in L$. Then x is almost integral over D if and only if there exist elements $a_1/b_1,\ldots,a_n/b_n \in L$ such that

$$D[x] \subseteq D \cdot (a_1/b_1) + \cdots + D \cdot (a_n/b_n)$$

If $d = b_1b_2\ldots b_n$, then $dx^k \in D$ for $k \geqslant 0$. Therefore, if x is almost integral over D, there exists a nonzero element $d \in D$ such that $dx^k \in D$ for $k \geqslant 0$. Conversely, if such an element d exists, then for $k \geqslant 0$, $x^k \in D \cdot (1/d)$ and so $D[x] \subseteq D \cdot (1/d)$. These remarks prove the following result.

THEOREM 1.15. Let D be an integral domain with quotient field L. An element $x \in L$ is almost integral over D if and only if there exists a nonzero element $d \in D$ such that $dx^k \in D$ for $k \geq 0$.

We are finally able to prove the result we promised sometime back. Recall that a *unique factorization domain* (UFD for short, and often also referred to as a *factorial domain*) is an integral domain with the property that each nonzero nonunit can be written uniquely as a product of prime elements. Well known examples include the rings Z, $Z[X_1,\ldots,X_n]$, and $L[X_1,\ldots,X_n]$ for L a field; see [22].

THEOREM 1.16. A unique factorization domain is completely integrally closed.

Proof. Let D be a UFD with quotient field L and let $x \in L$ be almost integral over D. By Theorem 1.15, there exists a a nonzero element $d \in D$ such that $dx^k \in D$ for $k \geq 0$. Since D is a UFD, we can write $x = a/b$ where a and b belong to D and have no prime factors in common. Thus, for $k \geq 0$, there exists an element $d_k \in D$ so that $d(a/b)^k = d_k$. Therefore, $da^k = d_k b^k$. Since a and b have no common prime factors, neither do a^k and b^k. It follows that if q is a prime factor of b, then q^k divides d for each $k \geq 0$. Again, because D is a UFD, this is impossible. Therefore, b has no prime factors and must be a unit of D. It follows that $x = a/b$ belongs to D and that D is completely integrally closed.

We close this section by stating without proof the following facts about an integral domain D. Let $X_1,\ldots,X_n$ be indeterminates.

The domain D is integrally closed if and only if $D[X_1,\ldots,X_n]$ is integrally closed [22, Theorem 10.7].

The domain D is completely integrally closed if and only if $D[X_1,\ldots,X_n]$ is completely integrally closed. [22, Theorem 13.6].

1.4 BÉZOUT DOMAINS

A *Bézout domain* is an integral domain D with the property that each finitely generated ideal of D is principal. Since each finitely generated ideal is principal if and only if each ideal with two generators is principal, the Bézout domains are precisely the domains with the property that each pair of elements has a greatest common divisor (g.c.d.) that is a linear combination of the elements. Here, by the g.c.d. of a and b we mean an element d such that d divides both a and b and any element that divides both a and b divides d. Since D is an integral domain, the g.c.d. of two elements is uniquely determined up to unit (that is, invertible element) multiples. This being so, we can write $(a,b) = (d)$, where (a,b) can mean either the ideal generated by a and b or the g.c.d. of a and b. Thus in a Bézout domain, we are justified in using the sometimes ambiguous notation $(a,b) = (d)$. (We say "ambiguous" because, for example, in $R[X,Y]$ the g.c.d. of X and Y is one, but $(X,Y) \neq (1)$.)

Obviously, any principal ideal domain (PID) is a Bézout domain. In fact, the PID's are exactly the noetherian Bézout domains. There are, of course, numerous important ring-theoretic examples of nonnoetherian Bézout domains and we shall give an important systems-theoretic example shortly. For now, we wish to prove that a Bézout domain is integrally closed. The following lemma will facilitate our proof.

LEMMA 1.17. Let D be a Bézout domain with $a,b \in D$. If n is a positive integer, then $(a,b) = (1)$ implies $(a^n,b) = (1)$. In words, if a and b are relatively prime, so are a^n and b.

Proof. We consider only the nontrivial case where both a and b are nonzero. If $(a^n,b) \neq (1) = D$, then there exists a maximal ideal M of D so that $(a^n,b) \subseteq M$. But then $a^n \in M$

and so $a \in M$. It follows that $(a,b) \subseteq M$ and that a and b are not relatively prime.

THEOREM 1.18. A Bézout domain is integrally closed.

Proof. Let D be a Bézout domain with quotient field L. If $a/b \in L$, we can assume that $(a,b) = (1)$. If a/b is integral over D, then there exists a positive integer n and ring elements $d_0, d_1, \ldots, d_{n-1} \in D$ so that

$$(a/b)^n + d_{n-1}(a/b)^{n-1} + \cdots + d_1(a/b) + d_0 = 0$$

From this it follows that

$$a^n = b(-d_{n-1}a^{n-1} - \cdots - d_1ab^{n-2} - d_0b^{n-1})$$

Therefore, b divides a^n. By Lemma 1.17, $(a^n,b) = (a,b) = (1)$, and hence, b is a unit of D. Therefore, $a/b \in D$ and D is integrally closed.

We remark that a Bézout domain need not be completely integrally closed. For example, any valuation domain of rank ≥ 2 provides an example [22, Theorem 17.5]. In the sequel we shall often be concerned with Bézout domains that are completely integrally closed. Principal ideal domains are such domains; another important example is given below.

EXAMPLE. Let R be the ring of real analytic functions defined on the real line. That is, R is the set of real-valued functions f, with domain the real line, having the following property: For each real number x_0, there exists an open neighborhood $V(f,x_0)$, of x_0 such that, for all x in $V(f,x_0)$, f(x) is the sum of an absolutely convergent power series in $x - x_0$. It follows immediately from the principle of isolated zeros, see [15, 9.1.5], that if $f \in R$, $f \neq 0$, then $\{x | f(x) = 0\}$ cannot have a cluster point. It is then immediate that R is an integral domain. In order to show that R is a Bézout domain, we need the following easy lemma.

LEMMA. Suppose R is a ring such that, for all a and b in R, there exists α, β, α', β', in R with $a\beta = \alpha b$ and $\alpha\alpha' + \beta\beta' = 1$. Then every finitely generated ideal of R is principal.

Proof. We show that every ideal of R generated by two elements is principal. Let a and b belong to R, and choose α, β, α', β' as in the statement of the lemma. Set $c = \alpha' a + \beta' b$. Then

$$\alpha c = \alpha\alpha' a + \beta'(\alpha b) = \alpha\alpha' a + \beta'(a\beta) = a$$

and

$$\beta c = \alpha'(\beta a) + \beta\beta' b = \alpha'(\alpha b) + \beta\beta' b = b$$

Hence $Ra + Rb \subseteq Rc \subseteq Ra + Rb$, and the doubly generated ideal $Ra + Rb$ is principal.

The following lemma then completes the proof of the fact that the ring of real analytic functions is a Bézout domain.

LEMMA. Let a and b belong to the ring R of real analytic functions. Then there exist elementa α and β in R such that $a\beta = \alpha b$ and $\alpha^2 + \beta^2 = 1$.

Proof. If $a = 0$, then $\beta = 1$ and $\alpha = 0$ have the desired properties. If neither a nor b is identically zero, then the zeros of both must be isolated. Let $t_1, t_2, t_3 \ldots$ be the zeros common to both a and b. Then by the Weierstrass factorization theorem, see [55, p. 323–325], a and b can be expanded as

$$a(t) = \bar{a}(t) \prod_{k=1}^{\infty} (1 - (t/t_k))^{n_k} e^{p_k(t)}$$

$$b(t) = \bar{b}(t) \prod_{k=1}^{\infty} (1 - (t/t_k))^{n_k} e^{q_k(t)}$$

where n_k is the order of the common zero t_k, p_k and q_k are convergence-producing polynomials, and $\bar{a}$ and $\bar{b}$ are real analytic functions with no common zeros. (The results of [55] are stated for complex analytic functions, but an examination of the proofs shows that the results we need hold for real analytic functions.) Define α and β by

$$\alpha(t) = \bar{a}(t)/(\bar{a}(t)^2 + \bar{b}(t)^2)^{1/2}$$
$$\beta(t) = \bar{b}(t)/(\bar{a}(t)^2 + \bar{b}(t)^2)^{1/2}$$

Since the composition of real analytic functions is real analytic, see [15, 9.3.2], α and β are in R. The elements α and β satisfy the conclusions of the lemma.

We now show that R is completely integrally closed. To this end, assume that f, g, and h are nonzero elements of R and $h(f/g)^n \in R$ for all $n \geq 1$. We must show that $(f/g) \in R$. First, we show that if $g(x_0) = 0$, then $f(x_0) = 0$. For $n \geq 1$, let $\theta_n \in R$ be such that $hf^n = g^n\theta_n$. If $f(x_0) \neq 0$, then $h(x_0) = 0$. Since $h \neq 0$, there is a neighborhood V of x_0 and an integer $k \geq 1$ such that

$$h(x) = (x - x_0)^k h_1(x)$$

for all $x \in V$, where $h_1 \in R$ and $h_1(x_0) \neq 0$. There is a neighborhood W of x_0 and an integer $\ell \geq 1$ such that

$$g(x) = (x - x_0)^\ell g_1(x)$$

for all $x \in W$, where $g_1 \in R$ and $g_1(x_0) \neq 0$. Choose n so large that $n\ell > k$. Then for x in $V \cap W$ we have

$$(x - x_0)^k h_1(x) f^n(x) = (x - x_0)^{n\ell} g_1^n(x)\theta_n(x)$$

Cancelling $(x - x_0)^k$, we see that $f(x_0) = 0$.

Now, assuming that $g(x_0) = 0$, let $g(x) = (x - x_0)g_1(x)$, where $g_1(x)$ is defined in a neighborhood W of x_0 and g_1 has a convergent power series valid in W. Then, by the above paragraph, $f(x) = (x - x_0)f_1(x)$ for some f_1 that has a convergent power series in some neighborhood V of x_0. Then $(f/g) = (f_1/g_1)$ in $V \cap W$. If $g_1(x_0) = 0$, repeat the process.

Eventually, we get $(f/g) = (f_m/g_m)$ in some neighborhood of x_0 with $g_m(x_0) \neq 0$. Hence, we can reduce to the case when $g(x_0) \neq 0$. In that case, (f/g) has a convergent power series in a neighborhood of x_0. So (f/g) is real analytic, and R is completely integrally closed. We summarize the last few pages in the following theorem.

THEOREM 1.19. The ring of real analytic functions defined on the real line is a completely integrally closed Bézout domain.

We turn our attention now to giving a homological characterization of Bézout domains. Let D be an integral domain. We claim that a nonzero ideal I of D is free as a D-module if and only if I is principal. To see this, if $I = (x)$, then $D \cong I$ under the map $d \longrightarrow dx$. Hence, I is a free D-module. Conversely, if I is free with basis $\{x_j\}$ and if there is more than one j, let x_1 and x_2 be distinct basis elements. Then $0 = x_2x_1 - x_1x_2$ in contradiction to the freeness of the x's. Therefore, $I = (x_1)$ is principal.

It follows that D is a Bézout domain if and only if each nonzero finitely generated ideal of D is free. In other words, D is a Bézout domain if and only if each finitely generated D-submodule of the free D-module D is free. Our next theorem extends this idea considerably.

THEOREM 1.20. Let D be an integral domain. Then D is a Bézout domain if and only if each finitely generated submodule of a free D-module is free.

Proof. If each finitely generated submodule of a free D-module is free, then each finitely generated ideal I of D is free. It follows that I is principal.

Conversely, if D is a Bézout domain, then each finitely generated ideal of D is a free D-module. Let A be a finitely generated submodule of the free D-module F. If $\{x_i\}$ is a basis for F, then since the finitely many generators of A involve only finitely many of the x's, it follows that $A \subseteq Dx_1 \oplus \cdots \oplus Dx_n$ for some positive integer n. To prove that A is free, we make an induction on n. If $n = 1$, then $A \subseteq Dx_1 \cong D$ and therefore A is isomorphic to a finitely generated ideal of D. For the induction step, consider the map $\phi : A \longrightarrow D$ defined as follows: Each element $a \in A$ can be written uniquely in the form

$$a = r_1x_1 + r_2x_2 + \cdots + r_nx_n$$

for elements $r_i \in D$. Set $\phi(a) = r_n$. Then ϕ is evidently a D-homomorphism and, consequently, $\phi(A)$ is a D-submodule of D—that is, an ideal of D. Since A is finitely generated, $\phi(A)$ is finitely generated and therefore free as a D-module. This all gives rise to the following exact sequence

$$0 \longrightarrow A \cap (Dx_1 \oplus \cdots \oplus Dx_{n-1}) \longrightarrow A \xrightarrow{\phi} \phi(A) \longrightarrow 0$$

where $\phi(A)$ is free and hence projective. Consequently, this sequence splits and we have that

$$A \cong \phi(A) \oplus [A \cap (Dx_1 \oplus \cdots \oplus Dx_{n-1})]$$

But $A \cap (Dx_1 \oplus \cdots \oplus Dx_{n-1})$ is finitely generated, being a homomorphic image of A, and is also a submodule of a free D-module on $(n - 1)$ generators. By the induction assumption, $A \cap (Dx_1 \oplus \cdots \oplus Dx_{n-1})$ is free and so is A. This completes the proof of Theorem 1.20.

We shall use Theorem 1.20 often in subsequent chapters. On at least one occasion we shall need a corollary to Theorem 1.20, but to prove it, we require the following fact. Recall that a module A over an integral domain D is called *torsion-free* if, for each $d \in D$, $a \in A$, from $da = 0$ it follows that $d = 0$ or $a = 0$. The result we need, but only state, is: A finitely generated torsion-free module over an integral domain

D is isomorphic to a submodule of a free D-module. (See [51, Lemma 4.31].)

THEOREM 1.21. Let D be an integral domain. Then D is a Bézout domain if and only if each finitely generated torsion-free D-module is free.

Proof. This is immediate from Theorem 1.20 and the result quoted above.

1.5. LOCALIZATION

In this brief section we sketch the essentials of the theory of localization. Since we will not be using local techniques very much in this book, a brief presentation should be sufficient.

Let R be a commutative ring with S a multiplicatively closed subset of elements that are not zero divisors of R. It is convenient to assume that $1 \in S$. Form the set R_S of formal fractions $\{r/s \mid r \in R, s \in S\}$ with equality given by $r/s = r'/s'$ if and only if $rs' = r's$. Add and multiply fractions in the usual way. Then R_S is a commutative ring with identity. Moreover, the mapping $r \longrightarrow r/1$ is an injective ring homomorphism from R into R_S and, consequently, we can assume that R is a subring of R_S. Notice that the principal effect of passing from R to R_S is to invert each element of S. Indeed, R_S is essentially the smallest ring containing R in which each element of S is a unit.

There are several important examples:

1. Let R be a ring with S the set of <u>all</u> elements of R that are not zero divisors of R. Then R_S is called the *total quotient ring of* R. In case D is an integral domain, then $S = D\setminus\{0\}$ and $D_S = L$, the quotient field of D.

2. Let D be an integral domain with P a prime ideal of D. The very definition of prime ideal is that $D\setminus P$ is multi-

plicatively closed. The standard notation here, and it is clearly abusive, is D_P instead of $D_{(D\backslash P)}$.

3. Let R be a commutative ring with s not a zero divisor of R. Then $S = \{s^i\}_{i=0}^{\infty}$ is multiplicatively closed. The standard notation here is R_s.

We must add that with care the restriction that S contain no zero divisors can be removed. Removing it allows one, for example, to treat (2) above for arbitrary rings. However, for our purposes no finesse is required and we will assume that S consists entirely of elements that are not zero divisors.

The following observation is often useful. Let X be an indeterminate over R with S a multiplicatively closed subset of R. Then S is a multiplicatively closed subset of the ring R[X] and it is easy to check that $(R[X])_S = R_S[X]$. In case D is an integral domain with quotient field L and $S = D\backslash\{0\}$, then $(D[X])_S = L[X]$.

Finally, we make the following observations about ideals of R_S. Since the elements of S are units of R_S, if I is an ideal of R with $I \cap S \neq \emptyset$, then $IR_S = R_S$. The converse also holds. Moreover, each ideal J of R_S is of the form $J = IR_S = \{i/s \mid i \in I, s \in S\}$ for some ideal I of R. In particular:

If R is noetherian, then R_S is noetherian.

1.6. PROJECTIVE MODULES AND INVERTIBLE IDEALS

In this section we expose some rudimentary properties of projective modules and invertible ideals in the hope of making Theorem 3.20 accessible. That goal has guided our choice of topics.

Let R be a commutative ring with E an R-module. Then E is said to be *projective* if and only if, whenever we are given a diagram of the form

$$\begin{array}{ccc} & & E \\ & & \downarrow\beta \\ B & \xrightarrow{\alpha} & C \longrightarrow 0 \end{array}$$

there exists an R-module homomorphism $\gamma : E \longrightarrow B$ such that $\alpha \circ \gamma = \beta$. It is well known and easy to see that E is projective if and only if each exact sequence of the form

$$0 \longrightarrow A \longrightarrow B \xrightarrow{\beta} E \longrightarrow 0$$

splits—that is, there exists a homomorphism $\gamma : E \longrightarrow B$ such that $\beta \circ \gamma = 1_E$, the identity on E. Not as well known, but just as important, is the following characterization.

THEOREM 1.22. (Dual Basis Lemma). An R-module E is projective if and only if there exists a subset $\{e_i\}_{i\in I}$ of E and R-homomorphisms $\{\phi_i : E \longrightarrow R\}_{i\in I}$ such that

1. For $e \in E$, $\phi_i(e) = 0$ for all but finitely many i;
2. For $e \in E$,

$$e = \sum_{i\in I} (\phi_i(e))e_i$$

Proof. Suppose that E is projective and let ψ be a surjection from a free R-module F onto E. Since E is projective, there is a map $\phi : E \longrightarrow F$ such that $\psi \circ \phi = 1_E$. Let $\{f_i\}_{i\in I}$ be a basis for the free module F. If $e \in E$, then the element $\phi(e)$ can be written uniquely in the form

$$\phi(e) = \sum_{i\in I} r_i f_i$$

for $r_i \in R$ and all but finitely many of the r's equal to zero. Define

$$\phi_i : E \longrightarrow R \quad \text{by} \quad \phi_i(e) = r_i$$

Clearly, $\phi_i(e) = 0$ for almost every i. Finally, define $e_i = \psi(f_i)$ for each $i \in I$. Then for $e \in E$, we have that

$$e = \psi(\phi(e)) = \psi(\sum_{i \in I} r_i f_i) = \sum_{i \in I} r_i \psi(f_i) = \sum_{i \in I} \phi(e_i) e_i$$

Therefore, both (1) and (2) hold.

Conversely, suppose that each of conditions (1) and (2) holds for the elements $\{e_i\}_{i \in I}$ and homomorphisms $\{\phi_i\}_{i \in I}$. Let F be a free R-module of rank equal to the cardinality of the set I and let $\{f_i\}_{i \in I}$ be a basis for F. Define a map $\psi : F \longrightarrow E$ by $\psi(f_i) = e_i$. This gives an exact sequence

$$F \xrightarrow{\psi} E \longrightarrow 0$$

If the sequence splits, then E is a direct summand of a free module and therefore is projective. Define $\phi : E \longrightarrow F$ by

$$\phi(e) = \sum_{i \in I} (\phi_i(e)) f_i \quad \text{for each} \quad e \in E$$

This map is well-defined by property (1). By property (2),

$$(\psi\phi)(e) = \psi(\sum_{i \in I} (\phi_i(e)) f_i) = \sum_{i \in I} (\phi_i(e)) \psi(f_i)$$

$$= \sum_{i \in I} (\phi_i(e)) e_i = e$$

Hence, $\psi \circ \phi = 1_E$ and the proof is complete.

Suppose that D is an integral domain with A an ideal of D. When is A a projective D-module? In order to answer this question, it is convenient to extend our definition of "ideal". Thus, let L be the quotient field of D. By a *(fractional) ideal* of D, we mean a D-submodule of L. It is customary to call ordinary ideals "integral ideals" to distinguish them from other types of fractional ideals.

Let A and B be fractional ideals of D. Since L is a ring, we can perform the usual ideal-theoretic operations on A and B. In particular, $A \cdot B$ is the D-submodule of L generated by the elements $a \cdot b$, where $a \in A$, $b \in B$. It is also a fractional ideal of D. Given a fractional ideal A of

D, we say that A is *invertible* if and only if there exists a fractional ideal B of D such that AB = D.

Perhaps, these ideas will become clearer if we look at a specific type of fractional ideal. Let d be a nonzero element of D. Then $(d) = \{xd \mid x \in D\}$ is a principal ideal of D. It is invertible with inverse $(1/d) = \{x/d \mid x \in D\}$. Thus, principal ideals are always invertible. In a PID, all nonzero ideals are invertible. Integral domains having the property that all nonzero ideals are invertible are called Dedekind domains, and we shall study these domains in the next section. However, for any integral domain D, we can consider the set I of all invertible (fractional) ideals of D. It is a group under the definition of product given above. Moreover, the domain D itself is the identity element of the group I. Note also that I always contains the collection of all principal (fractional) ideals of D. Furthermore, as a consequence of proving that I is a group, one has that an invertible ideal has a unique inverse. It can be shown that if A is an invertible ideal of D, then $A^{-1} = \{x \in L \mid xA \subseteq D\}$.

So, let A be an invertible ideal of D. Then $AA^{-1} = D$ and consequently there exist elements $x_1, \ldots, x_n \in A^{-1}$, $a_1, \ldots, a_n \in A$ such that $1 = x_1a_1 + \cdots + x_na_n$. Hence, if $a \in A$, then $a = a \cdot 1 = (ax_1)a_1 + \cdots + (ax_n)a_n$. Since $ax_i \in D$ for $1 \leqslant i \leqslant n$, this shows that the elements $a_1, \ldots, a_n$ generate A as an ideal of D.

We are at last ready to answer the question posed earlier.

THEOREM 1.23. Let D be an integral domain with quotient field L and let A be a nonzero ideal of D. Then A is a projective D-module if and only if A is an invertible ideal.

Proof. If A is projective, then by Theorem 1.22, there exist elements $\{a_i\}_{i\in I} \subseteq A$ and D-maps $\{\phi_i : A \to D\}_{i\in I}$ such that:

1. If $a \in A$, $\phi_i(a) = 0$ for almost all i; and
2. If $a \in A$, then

$$a = \sum_{i \in I} (\phi_i(a))a_i$$

Let $b \in A$, $b \neq 0$, and define $x_i \in L$ by $x_i = \phi_i(b)/b$. We prove that x_i is well-defined. If $b' \in A$ and $b' \neq 0$, then

$$b'\phi_i(b) = \phi_i(b'b) = \phi_i(bb') = b\phi_i(b')$$

so that $\phi_i(b)/b = \phi_i(b')/b'$. Thus, for each $i \in I$, we get a uniquely defined element $x_i \in L$.

We claim that $x_iA \subseteq D$ for all i. To see this, let $b \in A$, $b \neq 0$. Then $x_ib = [\phi_i(b)/b]b = \phi_i(b) \in D$. By condition (1), if $b \in A$, $b \neq 0$, then almost all $\phi_i(b) = 0$. Therefore, since $x_i = \phi_i(b)/b$, there are only finitely many x_i's. Moreover, by condition (2),

$$b = \sum_{i \in I} (\phi_i(b))a_i = \sum_{i \in I} (x_ib)a_i = b(\sum_{i \in I} x_ia_i)$$

Since $b \neq 0$, cancel b to obtain the equation $1 = \sum x_ia_i$. We have proved that there are only finitely many x_i's, that each one belongs to A^{-1}, and that $AA^{-1} = D$. It follows that A is invertible.

Conversely suppose that A is invertible and choose $a_1, \ldots, a_n \in A$ and $x_1, \ldots, x_n \in A^{-1}$ so that $1 = \sum_{i=1}^n x_ia_i$. For $1 \leqslant i \leqslant n$, define $\phi_i : A \to D$ by $\phi_i(a) = x_ia$. Note that $\phi_i(a) \in D$ since $x_i \in A^{-1}$. If $a \in A$, then

$$\sum(\phi_i(a))a_i = \sum(x_ia)a_i = a(\sum x_ia_i) = a \cdot 1 = a$$

Therefore, by the Dual Basis Lemma, Theorem 1.22, A is a projective D-module.

1.7 DEDEKIND DOMAINS

An integral domain D is called a *Dedekind domain* if and only if each nonzero ideal of D is invertible. While being an extremely important class of domains among those arising in number theory and algebraic geometry, Dedekind domains are rare as rings of functions. However, in this section we shall prove that the ring of real analytic functions on the unit circle is a Dedekind domain. Hopefully, this will be of interest to both algebraists and analysts. In Chapter Three we will discuss some control theory facts about this ring.

Let S^1 denote the unit circle and let R be the set of all real analytic functions defined on S^1. We recall what it means for a function to be real analytic on S^1: Consider the functions $e_0 : (0,2\pi) \longrightarrow \mathbf{R}$ and $e_1 : (-\pi,\pi) \longrightarrow \mathbf{R}$ defined by $e_0(x) = e^{ix}$ and $e_1(x) = e^{ix}$. For f to be real analytic on S^1 means that $f \circ e_0$ and $f \circ e_1$ are both real analytic; that is, have convergent Taylor's series expansions in a neighborhood of each point of their domains. It then follows from [15, 9.1.5] that if $f \in R$ is not identically zero, the zeros of f must be isolated. Since S^1 is compact, this implies that if $f \in R$ is not identically zero, then f has only finitely many zeros. Also, an element f of R which never vanishes is invertible in R (compose with the real analytic function 1/x and use [15, 9.3.2]). Under pointwise addition and multiplication, R is an integral domain.

For $\xi \in S^1$, let $M_\xi = \{f \in R | f(\xi) = 0\}$. We claim that $\{M_\xi\}_{\xi \in S^1}$ is the set of all maximal ideals of R. For $\xi \in S^1$, fixed, the map $\Phi_\xi : R \longrightarrow \mathbf{R}$ defined by $\Phi_\xi(f) = f(\xi)$ is a ring homomorphism of R onto $\mathbf{R}$, and consequently, the kernel M_ξ is a maximal ideal of R. Conversely, suppose that I is an ideal of R such that, for every point $\zeta \in S^1$, there exists an element $f \in I$ such that $f(\zeta) \neq 0$. Note that if f has a zero in each neighborhood of ζ, $f(\zeta) = 0$ by continuity. Hence, for each point $\xi \in S^1$, there is an element $f \in I$ and a neighborhood N_ξ of ξ such that $f(\xi)$ is never zero on N_ξ. Since S^1 is compact, this cover has a finite subcover $N_{\xi_1},\ldots,N_{\xi_s}$ with corresponding functions $f_1,\ldots,f_s \in I$. Let $f = f_1^2 + \cdots + f_s^2$. Then $f \in I$ and $f(\xi) \neq 0$ for each $\xi \in S^1$. As noted above, such a function must be a unit and hence $I = R$. This completes the proof that $\{M_\xi\}_{\xi\in S^1}$ is the set of all maximal ideals of R. Further, as noted above, an element f of R that is not identically zero has only finitely many zeros. Hence, each element of R belongs to only finitely many maximal ideals.

Now for $\xi \in S^1$ fixed, we define a function $v_\xi : R \longrightarrow Z$ as follows: Let

$$U_0 = e_0(0,2\pi) = S^1\backslash\{1\}$$

and

$$U_1 = e_1(-\pi,\pi) = S^1\backslash\{-1\}$$

If $h \in R$ and $\xi \in U_0$, there is a neighborhood N_0 of $\xi_0 = e_0^{-1}(\xi)$ such that $h \circ e_0$ has a convergent power series expansion valid on N_0. Thus, we can write

$$h \circ e_0 = a_0 + a_1(x - \xi_0) + a_2(x - \xi_0)^2 + \cdots$$

The least positive integer n such that $a_n \neq 0$ is called the *order of* h *at* ξ, and we set $v_\xi(h) = n$. If $\xi \in U_1$, there is a neighborhood N_1 of $\xi_1 = e_1^{-1}(\xi)$ such that $h \circ e_1$ has a convergent power series expansion valid on N_1. We let $v_\xi(h)$ be the order of the zero of $h \circ e_1$ at ξ_1. This definition of $v_\xi(h)$ is well-defined since $(e_0^{-1} \circ e_1)(\xi_1) = \xi_0$ and $e_0^{-1} \circ e_1$ is a real analytic isomorphism. Clearly the map v_ξ has the properties

1. $v_\xi(g \cdot h) = v_\xi(g) + v_\xi(h)$ and
2. $v_\xi(g + h) \geqslant \min\{v_\xi(g), v_\xi(h)\}$.

Such a function v defined on an integral domain D is called a *discrete valuation* on D, and it has an obvious extension to the quotient field L of D by defining $v(a/b) = v(a) - v(b)$. Then the subset $V = \{x \in L \mid v(x) \geq 0\}$ is a subdomain of L, called a *discrete valuation ring* (denoted DVR). A DVR is a principal ideal domain with only one maximal ideal. The interested reader is referred to [39, p. 67] or to [22, Chapter III].

Thus, in our case, for each $\xi \in S^1$, there is attached a discrete valuation v_ξ and a DVR

$$V_\xi = \{\frac{f}{g} \mid f,g \in R,\ g \neq 0,\ \text{and } v_\xi(f) \geq v_\xi(g)\}$$

We now show that

$$V_\xi = R_{M_\xi} = \{\frac{f}{g} \mid f,g \in R,\ g \notin M_\xi\}$$

If

$$\frac{f}{g} \in R_{M_\xi}$$

then we can assume that $g \in M_\xi$. Consequently, $v_\xi(g) = 0$ and $f/g \in V_\xi$. Conversely, let $f/g \in V_\xi$ and assume that $\xi \in U_0$ with $\xi = \exp(i\xi_0)$, $0 < \xi_0 < 2\pi$. First, assume that $v_\xi(g) = m$ is even and define h from S^1 to the reals by $h(e^{ix}) = \sin^m((x - \xi_0)/2)$. Since an even power of the sine is periodic of period π, h is well-defined and clearly real analytic. Since h is zero only at ξ and $v_\xi(h) = m = v_\xi(g) \leq v_\xi(f)$, it is clear that f/h and g/h are real analytic. Then $f/g = (f/h)/(g/h)$ and $(g/h)(\xi) \neq 0$, so that

$$\frac{f}{g} \in R_{M_\xi}$$

If $v_\xi(g)$ is odd, let $h(e^{ix}) = \sin(x - \xi_0)$. Then h is real analytic and $v_\xi(h) = 1$. Now write $f/g = (hf)/(hg)$ and note that $v_\xi(hf) - v_\xi(hg) = v_\xi(f) - v_\xi(g)$. An application of the even case then shows that

$$\frac{f}{g} \in R_{M_\xi}$$

This completes the proof of the converse.

After these preliminaries about the integral domain R, we prove a theorem from which it will follow that R is a Dedekind domain.

THEOREM 1.24. Let D be an integral domain with $\{M_i\}$ the collection of all maximal ideals of D. Suppose that D has the following two properties:

1. D_{M_i} is a DVR for each i, and
2. each nonzero element of D is contained in only finitely many maximal ideals of D.

Then D is a Dedekind domain.

Proof. We shall first prove that D is noetherian. Thus, let I be a nonzero ideal of D. Since each nonzero element of D is contained in only finitely many maximal ideals, the same is true for I. Let $M_1,\dots,M_2$ be the maximal ideals of D that contain I. Now, let a_0 be any nonzero element of I. Then the maximal ideals containing a_0 are $M_1,\dots,M_s$ together with (possibly) $N_1,\dots,N_t$. For $1 \leqslant j \leqslant t$, since $I \not\subseteq N_j$, there is an element $a_j \in I \backslash N_j$. Then $\{M_1,\dots,M_s\}$ is precisely the set of maximal ideals containing the elements $a_0,a_1,\dots,a_t$. Now, for $1 \leqslant k \leqslant s$,

since D_{M_k} is a DVR, ID_{M_k} is principal

generated by some element $b_k \in I$. We claim that $I = (a_0,a_1,\dots,a_t,b_1,\dots,b_s)$. Clearly, $(a_0,a_1,\dots,a_t,b_1,\dots,b_s) \subseteq I$. Now, it is an elementary exercise to prove that for S a commutative ring with ideals $A \supseteq B$, $A = B$ if and only if $AS_M = BS_M$ for each maximal ideal M of S. In our case, we have arranged precisely for that to happen by our selection of the a's and the b's. Consequently, I is finitely generated and D is noetherian.

So, to prove that D is a Dedekind domain, we have to prove that each nonzero ideal I of D is invertible. If $II^{-1} \neq D$, there exists a maximal ideal M of D such that $II^{-1} \subseteq M$. Since ID_M is principal, there exists an element $x \in I$ such that $xD_M = ID_M$. We also know that I is finitely generated, say $I = (a_1,\ldots,a_n)$. For $1 \leqslant i \leqslant n$, there exist elements $s_i \in D$, $s_i \notin M$ such that $s_ia_i \in xD$. Set $s = s_1\cdots s_n$. Then $(sx^{-1})a_i \in D$ for $1 \leqslant i \leqslant n$, and consequently, $sx^{-1} \in I^{-1}$. Therefore,

$$s = (sx^{-1})x \in I^{-1}I \subseteq M$$

a contradiction. It follows that $II^{-1} = D$, that I is invertible, and that D is a Dedekind domain.

We now record the promised result.

THEOREM 1.25. The ring of all real analytic functions defined on the unit circle is a Dedekind domain.

EXERCISES

1. Let R be a commutative ring with A an $n \times n$ matrix over R and B an $n \times m$ matrix over R. Let E be the R-submodule of R^n generated by the columns of the matrix $[B,AB,\ldots,A^{n-1}B]$. Prove: Let k be a positive integer, $k \geqslant n$. Then E is equal to the R-submodule of R^n generated by the columns of the matrix $[B,AB,\ldots,A^nB,A^{n+1}B,\ldots,A^kB]$.

2. Let **C**, **R**, and **Q** denote the fields of complex numbers, real numbers, and rational numbers, respectively. For R a ring let R[[X]] denote the ring of formal power series over R in the

indeterminate X. Let D be the following subring of $\mathbf{C}[[X]]$:

$$D = \{f \in \mathbf{C}[[X]] \mid f = \sum_{i=0}^{\infty} a_i X^i \text{ with } a_0 \in Q\}$$

Prove that D is not noetherian. (Hint: Show that the set of f in D with zero constant term is a nonfinitely generated ideal of D.)

3. Let R be the ring of all continuous functions on the interval [0,1]. Prove that R is not noetherian.

4. With the notation as in Exercise (2), prove that $\mathbf{C}[[X]]$ is integral over $\mathbf{R}[[X]]$. Is $\mathbf{C}[[X]]$ integral over $Q[[X]]$?

5. Let Z be the ring of integers and consider the polynomial ring $Z[X]$ in the indeterminate X. Let $S = Z\setminus\{0\}$. Prove that $Q[X] = (Z[X])_S$. Is $Q[[X]] = (Z[[X]])_S$?

6. Let $A \subseteq B$ be ideals of the integral domain D. Suppose that $AD_M = BD_M$ for each maximal ideal M of D. Prove: $A = B$. (Hint: If $b \in B\setminus A$, prove that $A:(b) = \{d \in D \mid db \in A\} = D$.)

7. Prove that a module E is projective if and only if each short exact sequence of the form

$$0 \longrightarrow A \longrightarrow B \longrightarrow E \longrightarrow 0$$

splits.

8. Assuming that A is an invertible ideal of D, prove that

$$A^{-1} = \{x \in L \mid xA \subseteq D\}$$

9. Let K be a field with v a discrete valuation on K. Thus, $v : K\setminus\{0\} \longrightarrow Z^+$ is surjective and satisfies $v(ab) = v(a) + v(b)$ and $v(a + b) \geqslant \min\{v(a), v(b)\}$. Set $V = \{x \in K \mid v(x) \geqslant 0\}$. Prove:

a. An element $x \in V$ is a unit if and only if $v(x) = 0$.

b. The ideal $M = \{x \in V | v(x) > 0\}$ is the unique maximal ideal of V.

c. Let $m \in M$ be such that $v(m) = 1$. Then $mV = M$.

d. If m is as in part (c), then $\{m^i V\}_{i=0}^{\infty}$ is the set of all ideals of V. In particular, V is a (local) principal ideal domain.

10. Consider the following subring of Q, the field of rational numbers.

$D = \{a/b | \ a, b \in Z, \ b \text{ odd}\}$

a. Prove that D is a DVR.

b. Prove that

$M = \{a/b | \ a \text{ is even}\}$

is the maximal ideal of D.

11. With the notation as in Exercise (2), prove that $\mathbf{C}[[X]]$ is a DVR whose maximal ideal is the set of all power series whose constant term is 0.

NOTES AND REMARKS

The material of the chapter is classical except for Theorem 1.25. Theorem 1.25 is from [25].

2
Reachability and Observability

In this chapter we shall introduce and study in detail the systems-theoretic notions of reachability and observability. We begin by giving the geometric definitions of reachability and observability when the ring R being considered is the real field **R**, and then state the classical theorems that relate the geometric definitions to the definitions that we will be using in this book. So consider the system

$$\frac{dx}{dt} = Fx(t) + Gu(t)$$
$$y(t) = Hx(t) \tag{1}$$

where x(t), y(t) and u(t) are column vectors, and F, G, and H are $n \times n$, $n \times m$, and $p \times n$ matrices, respectively, with entries in the field of real numbers **R**. This makes the *state* x(t) an n-vector, the *control* u(t) an m-vector and the *output* y(t) a p-vector. Such a system is called an *m-input, p-output, n-dimensional, continuous-time, time-invariant linear system* over **R** and can be designated by the triple (H,F,G). Usually we merely say that (H,F,G) is an n-dimensional system over **R** in this case, although the special cases when $m = 1$ and/or $p = 1$ are very important. The system is said to be a *single-input system* when $m = 1$ and a *single-output system* when $p = 1$.

Still considering the system (1), let the *initial state* $x(t_0) = x_0$ be given at time t_0. Then for a given control function $u(t)$, the state $x_u(t)$ of (1) at time $t > t_0$ is given by

$$x_u(t) = e^{F(t - t_0)}x_0 + \int_{t_0}^{t} e^{F(t - s)}Gu(s)\, ds$$

where $e^{F(t-t_0)}$ is the usual matrix exponential. The system (1) is said to be *reachable* if given x_0, t_0 and any $x_1 \in \mathbf{R}^n$, there is a time $t_1 > t_0$ and a control function $u(t)$ on $t_0 \leq t \leq t_1$ such that $x_u(t_1) = x_1$; that is, we can reach any point x_1 at some time t_1 starting from the initial point x_0 at time t_0. It is fairly easy to see that reachability for (1) does not depend on the initial point x_0, the initial time t_0, or the time $t_1 > t_0$. So, we henceforth assume that $x_0 = 0$ and $t_0 = 0$. For any matrix F, F^T denotes the transpose of F. The classical result about reachability is the following:

CLASSICAL REACHABILITY THEOREM. Let (H,F,G) be an n-dimensional system over $\mathbf{R}$. Then the following are equivalent.

1. (H,F,G) is reachable.
2. $G^T\exp(F^Ts)x = 0$ for $0 \leq s \leq t$, $t > 0$, implies $x = 0$.
3. $G^T(F^T)^jx = 0$ for each $j \geq 0$ implies $x = 0$.
4. The map $L_r: \mathbf{R}^{nm} \longrightarrow \mathbf{R}^n$ defined by $L_r = [G, FG, \ldots, F^{n-1}G]$ is surjective.
5. $\text{rank}\, L_r = \text{rank}[G, FG, \ldots, F^{n-1}G] = n$.

A similar result holds for observability, which we now define for the system (1) over $\mathbf{R}$. If we are given $t_0 = 0$ and the input $u(t)$ for $t \geq 0$, then the state $x_u(t)$ and output $y_u(t)$ are given by

$$x_u(t) = e^{Ft}x(0) + \int_0^t e^{F(t - s)}Gu(s)\, ds$$

and

$$y_u(t) = Hx_u(t) = He^{Ft}x(0) + \int_0^t He^{F(t-s)}Gu(s)\,ds$$

The system (1) is said to be *observable* if, given $t_0 = 0$, say, and $u(t)$ and $y_u(t)$ for $t \geq 0$, we can uniquely determine $x(0)$; that is, knowledge of the input and output uniquely determines the initial state of the system. Note that if $x(0)$ is uniquely determined, then so are all the states $x_u(t)$ for $t \geq 0$. Just as before, there is a classical result for observability.

CLASSICAL OBSERVABILITY THEOREM. Let (H,F,G) be an n-dimensional system over **R**. Then the following are equivalent.

1. (H,F,G) is observable.
2. $He^{Fs}x = 0$ for $0 \leq s \leq t$, $t > 0$, implies $x = 0$.
3. $HF^jx = 0$ for each $j \geq 0$ implies $x = 0$.
4. The map $L_o: \mathbf{R}^n \longrightarrow \mathbf{R}^{pn}$ defined by

$$L_o = \begin{bmatrix} H \\ HF \\ \vdots \\ HF^{n-1} \end{bmatrix}$$

is injective.

5. $\operatorname{rank} L_o = n$.

Proofs of these theorems can be found in most introductory texts on control theory; cf. [3]. It is also clear that these results are related in some way. In fact, each theorem follows from the other, if one first proves the following "duality theorem", which was observed in [30].

CLASSICAL DUALITY THEOREM. Consider the system (1) and its "dual" system (G^T,F^T,H^T) given by

$$z' = F^Tz(t) + H^Tu(t)$$
$$w(t) = G^Tz(t)$$

Then (H,F,G) is observable if and only if (G^T,F^T,H^T) is reachable.

We are now ready to consider systems (H,F,G) defined over a ring R and to define, in general, the concepts of reachability and observability. It is clear from the above classical theorems that there are several possible ways of defining these concepts. The definitions used here are the natural extensions to rings of parts (4) of the classical theorems above.

2.1 REACHABILITY

We start with a theorem about injectivity and surjectivity of homomorphisms of free modules.

THEOREM 2.1. Let R be a commutative ring and suppose that $\phi : R^m \longrightarrow R^n$ is a R-module homomorphism given by multiplication by some $n \times m$ matrix A.

1. Suppose that $m \geqslant n$ and let $I_n(A)$ be the ideal of R generated by the $n \times n$ minors of A. Then ϕ is surjective if and only if $I_n(A) = R$.

2. Suppose that $m \leqslant n$ and let $I_m(A)$ be the ideal of R generated by the $m \times m$ minors of A. Then ϕ is injective if and only if the annihilator of $I_m(A)$ is zero.

Proof. We begin by proving a result which is also of interest.

LEMMA 2.2. Let $A : E \longrightarrow F$ be a homomorphism of finitely generated R-modules. Then A is surjective if and only if A is residually surjective for each maximal ideal M of R—that is, the induced homomorphism $\bar{A} : E/ME \longrightarrow F/MF$ is surjective for all M.

Proof. Evidently, we have only to prove the sufficiency of the condition. This is one of the few places where we shall have need of "local methods". Let ImA denote the image of A. It suffices to prove that $F/\mathrm{Im}A = (0)$. We first show that for each maximal ideal M of R, $M(F/\mathrm{Im}A) = F/\mathrm{Im}A$.

Clearly, $M(F/\mathrm{Im}A) \subseteq F/\mathrm{Im}A$ and if $x \in F$, then by hypothesis, there exists an element $y \in E$ so that $x + MF = Ay + MF$, which means that $Ay - x = z$ for some $z \in MF$. But then $x + z \in \mathrm{Im}A$ and so $x + \mathrm{Im}A = -z + \mathrm{Im}A$ with $-z \in MF$.

Suppose for the moment that R has only one maximal ideal, say M. If F is a finitely generated module over R and if $MF = F$, then $F = (0)$. To see this weak version of Nakayama's Lemma, suppose that $x_1,\ldots,x_k$ is a minimal set of generators for F in the sense that no one of them can be deleted. Then

$$x_1 = m_1x_1 + m_2x_2 + \cdots + m_kx_k$$

for some $m_i \in M$ and so

$$(1 - m_1)x_1 = m_2x_2 + \cdots + m_kx_k$$

But M is the only maximal ideal of R and so $1 - m_1$ has an inverse in R. From this follows the contradiction that $x_1 \in \langle x_2,\ldots,x_k\rangle$.

In our case $F/\mathrm{Im}A$ is finitely generated and we saw in the first paragraph that $M(F/\mathrm{Im}A) = F/\mathrm{Im}A$. Thus, we would be finished if R has but one maximal ideal, and by localizing at M for every M, we can reach that situation. The result follows, for a module which is locally zero must be zero.

We have been too terse for some and too verbose for others. For the former, we suggest having a look at [2, Chapter 3].

Now for the proof of the theorem.

(1): From the lemma we have only to prove that ϕ is residually surjective for each maximal ideal M of R if and only if $I_n(A) = R$. But R/M is a field, called the residue class field of R at M, and hence ϕ is residually surjective at M if and only if the rank of $\bar{\phi}$ over R/M is n, which happens if and only if the rank of the matrix $\bar{A}$ over R/M is n. This is equivalent to saying that for each maximal ideal M of R, there is an $n \times n$ subdeterminant of A which is nonzero modulo M—that is, some $n \times n$ minor of A does not belong to M. This is clearly equivalent to saying that $I_n(A) = R$.

(2): Now, Φ fails to be injective if and only if there exists an element $x \in R^m \setminus (0)$ so that $\Phi(x) = 0$, which means that x is a nontrivial solution to the homogeneous system of linear equations $AX = 0$. Viewed this way, we have to prove

McCOY'S THEOREM. The homogeneous system of equations $AX = 0$ has a nontrivial solution if and only if the annihilator of the ideal $I_m(A)$ is nonzero.

Proof of McCoy's Theorem. Now A is an $n \times n$ matrix, say $A = (a_{ij})$. Suppose that $(b_1, \ldots, b_m)$ is a nontrivial solution to $AX = 0$ with $b_k \neq 0$. Let d be a determinant of order m which can be formed from A and, for the sake of notation, suppose that its elements come from the first m rows of A. Then

$$\sum_{j=1}^{m} a_{ij} b_j = 0 \text{ for } 1 \leq i \leq n$$

If we multiply the first of these equations by the (1,k)-cofactor in d and, in general, the ith equation by the (i,k)-cofactor in d, and add for $1 \leq i \leq m$, we have that $b_k \cdot d = 0$. The reason is that the sum of the products of the elements of any column (or row) of a determinant by the cofactors of the corresponding elements of a different column (or row) is always 0. A similar calculation shows that b_k annihilates each subdeterminant of A of size m and hence that b_k annihilates $I_m(A)$.

Conversely, suppose that there is a nonzero element b of R annihilating $I_m(A)$. If b annihilates each element of A, then $(b, b, \ldots, b)$ is a nontrivial solution of $AX = 0$. Otherwise, choose r maximal such that there exists an $r \times r$ submatrix B of A such that $b \cdot \det(B) \neq 0$. Suppose also that $r < n$, which we can do by adding the zero equation if necessary. Moreover, for convenience, suppose that B is in the upper left hand corner of A. Now consider the $(r + 1) \times (r + 1)$ submatrix of A located in the upper left hand corner and let $c_1, c_2, \ldots, c_{r+1}$ be the respective cofactors of the last row. Note that c_{r+1} equals $\det(B)$. We claim that

$$x_i = bc_i \text{ for } 1 \leq i \leq r + 1 \text{ and}$$

$$x_i = 0 \text{ for } r + 2 \leq i \leq m$$

is a nontrivial solution. It is certainly nontrivial, for $x_{r+1} \neq 0$ by choice. To check that it actually is a solution, merely substitute into $AX = 0$. The first r equations are satisfied for, as noted in the first paragraph, the sum of the products of the elements of any row of a determinant by the cofactors of the corresponding elements of another row is zero. As for the remaining equations, they are satisfied since b annihilates any minor of A of order $r + 1$. This completes the proof of the theorem.

We are now ready to define the notion of reachability for commutative rings.

Let (H,F,G) be an n-dimensional system over a ring R. We define the reachability map ρ from $\oplus_{i=0}^{\infty} R^m$ into R^n as follows: For each nonnegative integer i, $F^i G$ is an R-linear map from R^m to R^n. Set

$$\rho = \bigoplus_{i=0}^{\infty} F^i G$$

Thus, ρ is the direct sum map on $\oplus_{i=0}^{\infty} R^m$ and the matrix representation of ρ in some basis is $[G, FG, F^2 G, \ldots]$. We say that the system is *reachable* if and only if ρ is surjective, and we have at once the following theorem.

THEOREM 2.3. Let (H,F,G) be an n-dimensional system over a commutative ring R. The following are equivalent.

1. The system (H,F,G) is reachable.
2. The columns of $[G,FG,\ldots]$ generate R^n.
3. The columns of $A = [G,FG,\ldots,F^{n-1}G]$ generate R^n.

If ϕ denotes the map from $R^{mn} \longrightarrow R^n$ given by multiplication by A, then (1)-(3) are equivalent to

4. The mapping ϕ is residually surjective at each maximal ideal of R.
5. The ideal $I_n(A)$ generated by the $n \times n$ minors of A equals R.

Proof. (1) ⟹ (2): Let $x \in R^n$. There exists a positive integer N and an element $u = [u_0,\ldots,u_{N-1}] \in (R^m)^N$ such that

$$\rho(u) = \sum_{i=0}^{N-1} F^i G u_i = x$$

But, for $0 \leqslant i \leqslant N - 1$, $F^i G u_i$ is a linear combination of the columns of $F^i G$ and it follows that the columns of $[G,FG,\ldots]$ generate R^n.

(2) ⟹ (3): By the Cayley-Hamilton theorem, valid over any commutative ring (see Theorem 1.1), there exist elements $a_0,a_1,\ldots,a_{n-1} \in R$ such that

$$F^n = a_0 + a_1 F + \cdots + a_{n-1} F^{n-1}$$

Thus, for all $j \geqslant n$, $F^j \in R + RF + \cdots + RF^{n-1}$—that is to say, all powers of F belong to the R-module generated by 1, $F,\ldots,F^{n-1}$. Therefore, for $j \leqslant n$,

$$F^j G \in RG + RFG + \cdots + RF^{n-1}G$$

It follows that each column of $F^j G$ is a linear combination of the columns of $RG + \cdots + RF^{n-1}G$ and hence that the column modules of $[G,FG,\ldots]$ and $[G,FG,\ldots,F^{n-1}G]$ are identical.

(3) ⟹ (1): If $x \in R^n$, there exists an element $u \in R^{mn}$ such that $Au = \rho(u) = x$.

(3) ⟺ (4): Notice that the R-module generated by the columns of the matrix $[G,FG,\ldots,F^{n-1}G]$ is precisely the image of the mappings ϕ and ρ. Thus, the system (H,F,G) is reachable if and only if ϕ is surjective which, by Lemma 2.2, holds if and only if ϕ is residually surjective.

(1) ⟺ (5): By Theorem 2.1, part (1), ϕ is surjective if and only if $I_n(A) = R$.

REMARKS. In case R is a field, reachability of the system (H,F,G) is precisely the condition that the matrix $[G,FG,\ldots, F^{n-1}G]$ has rank n. Over any ring, if $m = 1$, reachability of (H,F,G) is, by part (5), equivalent to invertibility of the matrix $[G,FG,\ldots,F^{n-1}G]$.

Obviously, the reachability of a system depends only on the state matrix F and the input matrix G. We shall sometimes speak of the system (F,G) as being reachable.

In practice one often knows that a system (H,F,G) is reachable, and then another question arises. Given a vector $x \in R^n$, can we find a vector $u \in R^{mn}$ such that $\phi(u) = x$, and if so, how? This is obviously a difficult problem, but a result in this direction will be available to us after some preliminaries.

Recall that a field L is called *formally real* if -1 is not a sum of squares in L. The field of real numbers is formally real while the field of complex numbers is not.

THEOREM 2.4. Let L be a formally real field with A a matrix over L. Then A and A^TA have the same rank. In particular, applying this remark to A^T, we see that A^T (and hence A) and AA^T have the same rank.

Proof. We regard A and A^TA as being linear transformations on the appropriate L-vector spaces and we prove that null space A equals null space A^TA. The result will follow from the rank-nullity theorem. If $x \in$ null space A, then $x \in$ null space A^TA. Conversely, if $(A^TA)x = 0$, then $x^T(A^TA)x = (Ax)^TAx = 0$. But $(Ax)^TAx$ is a sum of squares and cannot be zero unless each term is zero, because L is formally real. Hence, Ax is zero as desired.

THEOREM 2.5. Suppose that R is a commutative ring with the property that each of its residue class fields is formally real. Let (H,F,G) be an n-dimensional system over R and set $A = [G,FG,\ldots,F^{n-1}G]$. Then (H,F,G) is reachable if and only if AA^T is invertible. If (H,F,G) is reachable and if $x \in R^n$, then the equation $Au = x$ has the solution $u = A^T(AA^T)^{-1}x$.

Proof. By Theorem 2.3, (H,F,G) is reachable if and only if Φ is residually reachable for each maximal ideal M of R. If M is a maximal ideal of R, then R/M is a formally real field, and Φ is residually surjective at M if and only if $\mathrm{rank}_{R/M}(M(A)) = n$, where M(A) denotes the matrix obtained from A by reducing coefficients modulo M. By Theorem 2.4, $\mathrm{rank}_{R/M}(M(A)) = n$ if and only if $\mathrm{rank}_{R/M}(M(A)[M(A)]^T) = n$. Since $M(A)[M(A)]^T$ is an $n \times n$ matrix, this condition holds if and only if $M(A)[M(A)]^T$ is invertible over R/M, which means that $\det(M(A)[M(A)]^T) \neq 0$ for each maximal ideal M of R. But this is the same as saying that $\det(AA^T)$ is a unit of R, which is equivalent to the invertibitliy of AA^T. To see the equivalence, one may use Cramer's Rule exactly as in the field case.

Theorem 2.5 is of course applicable to any formally real field as well as to certain algebras over such fields, but not to polynomial rings over them. Indeed, if K is any formally real field, then $(X^2 + 1)$ is a maximal ideal of K[X] and hence

$K[X]/(X^2 + 1)$ is a non-formally real residue field of $K[X]$. More concretely, taking $K = \mathbf{R}$, we are led to the complex field.

Theorem 2.5 fails in general, for consider the ring Z of integers. Let $G : Z^2 \to Z$ be given by [1 1] and take F and H to be as you like. Then A = G = [1 1] and so the columns of G generate Z, but $AA^T = [2]$ and 2 is not a unit of Z. Thus, (H,F,G) is reachable, while AA^T is not invertible.

2.2 OBSERVABILITY

Let R be a commutative ring with (H,F,G) a system over R. We wish now to define a notion that in some sense is dual to reachability. Reachability was defined in terms of some map's being surjective, while observability is defined in terms of a certain map's being injective.

Consider the R-homomorphism

$$\tau: R^n \to \prod_{i=0}^{\infty} R^p$$

defined by $\tau(x) = (Hx, HFx, HF^2x, \ldots)$. We say that the system (H,F,G) is *observable* if the map τ is injective, and we immediately give a characterization.

THEOREM 2.6. Let R be a commutative ring with (H,F,G) a system over R. The following are equivalent.

1. The system (H,F,G) is observable.
2. The R-homomorphism

$$\tau_n : R^n \to \prod_{i=0}^{n-1} R^p$$

given by $\tau_n(x) = (Hx, HFx, \ldots, HF^{n-1}x)$ is injective.

3. Set $B = [H^T, F^TH^T, \ldots, (F^T)^{n-1}H^T]$. If $I_n(B)$ denotes the ideal of R generated by the $n \times n$ minors of B, then the annihilator of $I_n(B)$ is zero.

Proof. (1) <==> (2): If τ is injective and if $\tau_n(x) = 0$, then $\tau(x) = 0$ by the Cayley-Hamilton theorem. Conversely, if τ_n is injective and if $\tau(x) = 0$, then $\tau_n(x) = 0$ and so $x = 0$.

(2) <==> (3): This follows immediately from Theorem 2.1, part (2).

Notice that Theorems 2.3 and 2.6 are not dual to one another. What is missing is the fact that a map can be injective without being residually injective. For example, in the ring Z of integers, multiplication by 2 is injective, but not residually injective at the maximal ideal (2). It is interesting that a residually injective homomorphism must be injective. This fact, however, we shall neither need nor prove.

We suggested earlier that observability was in a sense dual to reachability. To be more precise, if (H,F,G) is a system over a commutative ring R, then the *dual* of (H,F,G) is the system determined by the triple (G^T, F^T, H^T). Diagramatically, from

$$R^m \xrightarrow{G} R^n \xrightarrow{F} R^n \xrightarrow{H} R^p$$

we pass to

$$R^p \xrightarrow{H^T} R^n \xrightarrow{F^T} R^n \xrightarrow{G^T} R^m$$

It is classical that if R is a field, then a system (H,F,G) is reachable (resp. observable) if and only if the dual system (G^T, F^T, H^T) is observable (resp., reachable). We aim now at ring-theoretic extensions of these results.

THEOREM 2.7. Let R be a commutative ring with (H,F,G) an n-dimensional system over R. If the dual system (G^T, F^T, H^T) is reachable, then the system (H,F,G) is observable.

Proof. Since the system (G^T, F^T, H^T) is reachable, if $B = [H^T, F^TH^T, \ldots, (F^T)^{n-1}H^T]$, then $I_n(B) = R$ by Theorem 2.3. *A fortiori*, the annihilator of $I_n(B) = 0$ and so the system (H, F, G) is observable by Theorem 2.6.

Before stating our main duality theorem we need one additional fact.

LEMMA 2.8. Let R be a commutative ring with total quotient ring T and let (H,F,G) be a system over R. Then the system (H,F,G) is observable over R if and only if it is observable over T.

Proof. Now

$$T = \left\{\frac{a}{s} \;\middle|\; a, s \in R \text{ with } s \text{ not a zero divisor}\right\}$$

where fractions are added and multiplied as expected. We regard R as being a subring of T and so we are essentially, in our case, dealing with two maps $\tau_n : R^n \longrightarrow R^{pn}$ and $\tau_n^* : T^n \longrightarrow T^{pn}$. Since $\tau_n = \tau_n^*|_R$, if τ_n^* is injective, so is τ_n. If τ_n is injective and if $\tau_n^*(x) = 0$ for some $x \in T^n$, then we can, by finding a common denominator, write $x = (1/s)x^*$ for some $s \in R$ that is not a zero divisor and some $x^* \in R^n$. Then

$$0 = \tau_n^*(x) = \left(\frac{1}{s}\right)\tau_n^*(x^*) = \left(\frac{1}{s}\right)\tau_n(x^*)$$

Thus, $\tau_n(x^*) = 0$ which implies that $x^* = 0$. From $x^* = 0$ it follows that $x = 0$ and that τ_n^* is injective.

THEOREM 2.9. (The Duality Theorem). Let R be a noetherian ring with total quotient ring T and let (H,F,G) be an n-dimensional system over R. Set $B = [H^T, F^TH^T, \ldots, (F^T)^{n-1}H^T]$. The following are equivalent.

1. The system (H,F,G) is observable over R.
2. The ideal $I_n(B)$ generated by the $n \times n$ minors of B contains an element that is not a zero divisor.

3. The dual system (G^T,F^T,H^T) is reachable over T.
Thus, if R is a noetherian ring that is equal to its own total quotient ring, then the system (H,F,G) is reachable (resp. observable) if and only if the dual system (G^T,F^T,H^T) is observable (resp. reachable).

Proof. (1) ==> (2): It is an important property of noetherian rings that in such rings an ideal consisting entirely of zero divisors has a nonzero annihilator (Chapter 1, Theorem 1.8). Thus, by Theorem 2.6, part (3), $I_n(B)$ contains an element that is not a zero divisor.

(2) ==> (1): It follows again from Theorem 2.6 that condition (2) is formally stronger than condition (1).

(2) ==> (3): If condition (2) holds, then $[I_n(B)]T = T$, where $[I_n(B)]T$ denotes the ideal of T generated by the $n \times n$ minors of the matrix B. Since $[I_n(B)]T$ is the ideal over T of the defining matrix of the map $\tau_n{}^*$ defined in the proof of Lemma 2.8, it follows from Theorem 2.3 that the dual system (G^T,F^T,H^T) is reachable over T.

(3) ==> (1): The dual system (G^T,F^T,H^T) is reachable over T and hence, by Theorem 2.7, the system (H,F,G) is observable over T. By Lemma 2.8, the system (H,F,G) is observable over R.

REMARKS. Note that only in the proof of the implication (1) ==> (2) was the fact that R is noetherian used.

Also, a direct sum of fields satisfies the hypotheses of Theorem 2.9 and consequently the duality theorem holds for such rings. This observation recovers the Classical Duality Theorem for fields and also shows that it applies to rings such as those in [41, p. 19].

The duality theorem is not true without some hypothesis on the ring as the following example shows.

EXAMPLE. This is an example of an observable system over a noetherian ring whose dual is not reachable. Let Z be the ring of integers (whose total quotient ring is the field of rational numbers). Consider the one-dimensional system determined by F = [1], H = [2], and G arbitrary. Here, τ_n is multiplication by 2, clearly an injective mapping on Z. But the reachability map on Z of the dual system (G^T, F^T, H^T) is also multiplication by 2, which is not surjective. Hence the dual system is not reachable.

Changing directions, we seek an analogue to Theorem 2.5. Because injectivity does not imply residual injectivity, the following theorem is the best analogue available.

THEOREM 2.10. Let K be a formally real field with (H,F,G) a system over K and set $B = [H^T, F^T H^T, \ldots, (F^T)^{n-1} H^T]$. Then (H, F,G) is observable if and only if BB^T is invertible. Moreover, if (H,F,G) is observable and if $y_0, y_1, \ldots, y_{n-1} \in R^p$, then the equation

$$B^T x = \begin{bmatrix} y_0 \\ y_1 \\ \vdots \\ y_{n-1} \end{bmatrix}$$

has the solution

$$x = (BB^T)^{-1} B \begin{bmatrix} y_0 \\ y_1 \\ \vdots \\ y_{n-1} \end{bmatrix}$$

Proof. By Theorem 2.9, the system (H,F,G) is observable if and only if the dual system (G^T, F^T, H^T) is reachable if and only if BB^T is invertible by Theorem 2.5.

As for the moreover assertion, if $B^T x = y$, then $By = BB^T x$. Thus,

$$(BB^T)^{-1}By = (BB^T)^{-1}(BB^T)x = x$$

EXERCISES

1. Prove that the system (H,F,G) in Exercise (3) of Chapter 0 is both observable and reachable. (Hint: Use Theorem 2.3, part (3).)

2. a. Let (H,F,G) be an n-dimensional system over the real field **R**. Prove the Classical Duality Theorem: The system (H,F,G) is observable if and only if its dual system (G^T, F^T, H^T) is reachable.
b. Show that the Classical Duality Theorem also holds for systems (H,F,G) defined over the complex field **C**; i.e. show that (H,F,G) is observable if and only if (G^*, F^*, H^*) is reachable, where A^* denotes the conjugate transpose of a matrix A.
c. Let (H,F,G) be an n-dimensional system over an arbitrary field R. Show that (H,F,G) is observable if and only if the dual system (G^T, F^T, H^T) is reachable.

3. Let (F,G) be an n-dimensional system over a commutative ring R. Prove that (F,G) is reachable if and only if the system

$$\left(\begin{bmatrix} F & 0 \\ 0 & 0 \end{bmatrix}, \begin{bmatrix} G & 0 \\ 0 & I \end{bmatrix}\right)$$

is reachable, where I is the $r \times r$ identity matrix, $r \geq 0$.

4. Let (H,F,G) be an n-dimensional system over a commutative ring R. If (H,F,G) is reachable, prove that each of the following systems is also reachable over R.

a. $(HA, A^{-1}FA, A^{-1}G)$, where A is an invertible matrix.

b. (H, F, GB), where B is an invertible matrix.

c. $(H, F + I, G)$, where I is the $n \times n$ identity matrix.

d. $(H, F + GK, G)$, where K is an arbitrary matrix of the appropriate size.

5. Let

$$A = \begin{bmatrix} 1 & i \\ -i & 1 \end{bmatrix}$$

considered over the complexes. Prove that $AA^T = 0$, and hence that the hypothesis of formally real is necessary in Theorem 2.4.

6. Show directly that the ring of integers **Z** does not have the property that **Z**/M is formally real for every maximal ideal M of **Z**.

7. Let $R = \mathbf{R}[x,y]$. Consider

$$F = \begin{bmatrix} 1 & -1 \\ 1 & 1 \end{bmatrix}$$

$$G = \begin{bmatrix} y + x & y + x - 1 + x^2 + y^2 \\ y - x & y - x + 1 - x^2 - y^2 \end{bmatrix}$$

Prove that (F,G) is reachable. (Hint: Use Theorem 2.3, part (5).)

8. Let $R = Z[x]$, where Z denotes the integers. Let

$$F = \begin{bmatrix} 1 & 2 \\ 1 & 1 \end{bmatrix}$$

$$G = \begin{bmatrix} x - 2 & 3 \\ -3 & x + 2 \end{bmatrix}$$

Prove that (F,G) is reachable. (Hint: Use Theorem 2.3, part (5).)

NOTES AND REMARKS

Theorem 2.3 is the standard definition of reachability, as in [33] or [59], combined with facts from commutative algebra. Theorem 2.5, in the case that R itself is a formally real field, is Theorem 2.5 of [33]. The counterexample to Theorem 2.5 for arbitrary rings is from [33], Example 3.6]. Theorem 2.6 combines the standard definition of observability, as in [33] or [59], with facts from commutative algebra.

Theorem 2.9, the duality theorem, is in [14, Theorem 3.4]. The classical case of Theorem 2.9 (R a field) is in [30]. Our Theorem 2.10 is from [33, Theorem 3.10].

The systems in Exercises (7) and (8) appeared in [10] as examples to show that R[X,Y] and Z(X) do not have the pole assignability property.

3
Pole Assignability and Stabilizability

In this chapter we shall discuss the possibility of adjusting the eigenvalues of a system by "state feedback." For example, suppose that R is a ring of complex-valued functions on a set S. Let (H,F,G) be a system defined over R. For $s \in S$, let (H(s),F(s),G(s)) be the system over the complexes defined by evaluating the entries of the matrices at s. The stabilizability problem is the problem of determining when we can find a feedback matrix K over R such that for each element s in S the eigenvalues of the matrix $F(s) - G(s) \cdot K(s)$ are all contained in the open left half-plane. Phrased in this way, the problem makes no sense for systems defined over an arbitrary commutative ring. However, a related problem, the pole assignability problem, can be stated for systems defined over an arbitrary commutative ring and leads to interesting commutative algebra.

For the reader who wishes to see the "black box" formulation of state feedback, we have given it in Section 3.4. There one will also find the black box description of dynamic feedback. Although these can be omitted, we believe that a better understanding is achieved by reading them.

We shall study pole assignability first, returning to stabilizability in Section 3.5. Roughly speaking, the pole assignability problem is the following. Suppose that (H,F,G) is a reachable system over a commutative ring R. Can we, by means of state feedback, arbitrarily assign the eigenvalues (= "poles") of the system? More precisely, suppose that (H,F,G) is a reachable system over R of dimension n. If $r_1,\ldots,r_n \in R$ are given, can we find a matrix K such that the characteristic polynomial of F - GK is $(X - r_1)\cdots(X - r_n)$? In case this is possible for the system (H,F,G), we shall say that the system is *pole assignable*. If the ring R is such that all reachable systems over R are pole assignable, then we shall say that R has the *pole assignability property*. A stronger property with which we shall sometimes be concerned is defined as follows. Suppose that (H,F,G) is a reachable system over R of dimension n. If $a_0 + a_1X + \cdots + a_{n-1}X^{n-1} + X^n$ is a monic polynomial over R, can we find a matrix K over R such that the characteristic polynomial of F - GK is $a_0 + a_1X + \cdots + a_{n-1}X^{n-1} + X^n$? In case this is possible for the system (H,F,G), we shall say that the system is *coefficient assignable*. If the ring R is such that all reachable systems over R are coefficient assignable, then we shall say that R has the *coefficient assignability property*. It is obvious that if R has the coefficient assignability property, then R has the pole assignability property.

We begin with a result which tells why we are only interested in this question for reachable systems. Note that since the homomorphism H plays no role in reachability, we will throughout this chapter write our systems as pairs (F,G).

THEOREM 3.1. Let R be a commutative ring with (F,G) a system over R. If (F,G) is pole assignable, then (F,G) is reachable.

Proof. By Theorem 2.3, the system (F,G) is reachable if and only if it is residually reachable for each maximal ideal

M of R. Let $(\bar{F},\bar{G})$ be the residual system at M. If $\bar{r}_1,\ldots,\bar{r}_n \in R/M$, then there exists K such that

$$\det(XI - (F - GK)) = (X - r_1)\cdots(X - r_n)$$

and so

$$\det(X\bar{I} - (\bar{F} - \bar{G}\bar{K})) = (X - \bar{r}_1)\cdots(X - \bar{r}_n)$$

Thus, the residual system is pole assignable over the field R/M. This reduces our problem to the case when R is a field. So, start over and assume that R is a field.

Now F is an $n \times n$ matrix and G is an $n \times m$ matrix. Let V be the subspace of R^n generated by the columns of the matrix $[G,FG,\ldots,F^{n-1}G]$ and suppose that $V \neq R^n$. Then V is invariant under F, and for any $x \in R^n$ and any $m \times n$ matrix K, $(GK)x \in V$. Therefore, F and F − GK induce the same homomorphism, say F_1, on the quotient space R^n/V. Choose a basis $\{v_1,\ldots,v_k\}$ for V and extend it to a basis B of R^n. For any $m \times n$ matrix K, V is invariant under F − GK, and therefore,

$$M = [F - GK]_B = \begin{bmatrix} M_1 & M_2 \\ 0 & M_3 \end{bmatrix} \text{ for some } M_1,\ M_2, \text{ and } M_3$$

Now M_3 is the matrix of F_1 with respect to the basis $\{\bar{v}_{k+1}, \ldots,\bar{v}_n\}$ of R^n/V. Hence the characteristic polynomial of F − GK is

$$\begin{aligned} \det(XI_n - M) &= [\det(XI_k - M_1)][\det(XI_{n-k} - M_3)] \\ &= [\det(XI_k - M_1)] \cdot [\text{char. poly. of } F_1] \end{aligned}$$

Thus, for any K, the characteristic polynomial f of F_1 divides the characteristic polynomial of F − GK, although F_1 depends only on F and G. Now, the polynomials X^n and $(X - 1)^n$ are relatively prime, and since we can assign poles to (F,G), we must have that f divides both X^n and $(X - 1)^n$. It follows that f is a constant, which contradicts the fact that $V \neq R^n$. This completes the proof.

3.1 THE SINGLE-INPUT CASE AND FEEDBACK CYCLIZATION

We shall give examples in Section 3.3 to show that the concepts of reachability and pole assignability are distinct. More precisely, there exists a ring R and a reachable system (F,G) over R such that the system (F,G) is not pole assignable. Consequently, it is meaningful to try to determine when reachable systems are pole assignable. We shall pursue this in two directions and begin by making assumptions on G rather than on R. We require some notation.

If G consists of a single column vector, then the system (F,G) is called a *single-input* system. It is customary in this case to write (F,g) in place of (F,G). Notice that if (F,g) is n-dimensional, then (F,g) is reachable if and only if $\det([g, Fg, \ldots, F^{n-1}g])$ is a unit of R. (See Theorem 2.3.)

Let R be a commutative ring with $f \in R[X]$, say $f = X^n + a_{n-1}X^{n-1} + \cdots + a_1X + a_0$. There are two matrices naturally associated with f which are called "companion" matrices for f. These are

$$\bar{F} = \begin{bmatrix} 0 & 0 & 0 & \cdot & \cdot & \cdot & -a_0 \\ 1 & 0 & 0 & \cdot & \cdot & \cdot & -a_1 \\ 0 & 1 & 0 & \cdot & \cdot & \cdot & -a_2 \\ \vdots & & & & & & \\ 0 & 0 & 0 & \cdot & \cdot & \cdot & -a_{n-1} \end{bmatrix}$$

which we call the *companion matrix* of f, and

$$\hat{F} = \begin{bmatrix} 0 & 1 & 0 & . & . & . & 0 \\ 0 & 0 & 1 & . & . & . & 0 \\ 0 & 0 & 0 & . & . & . & 0 \\ \vdots & & & & & & \\ -a_0 & -a_1 & -a_2 & . & . & . & -a_{n-1} \end{bmatrix}$$

which we call the *alternate companion matrix* of f. It is easy to see that f is the characteristic polynomial of each of these matrices.

Let us make one additional definition. For an $n \times n$ matrix F over R, in analogy with the field case, we will say that a vector $v \in R^n$ is a *cyclic vector for* F if the set $\{v, Fv, \ldots, F^{n-1}v\}$ is a basis for R^n.

With this notation in hand, we now state the theorem in the single-input case.

THEOREM 3.2. Let (F,g) be a single-input n-dimensional system over a commutative ring R and let f be the characteristic polynomial for F. The following are equivalent.

1. The system (F,g) is reachable.
2. The vector g is a cyclic vector for F.
3. The matrix $[g, Fg, \ldots, F^{n-1}g]^{-1}F[g, Fg, \ldots, F^{n-1}g]$ is the companion matrix $\overline{F}$ for f.
4. The matrix $[\hat{g}, \hat{F}\hat{g}, \ldots, (\hat{F})^{n-1}\hat{g}][g, Fg, \ldots, F^{n-1}g]^{-1} \cdot F[g, Fg, \ldots, F^{n-1}g][\hat{g}, \hat{F}\hat{g}, \ldots, (\hat{F})^{n-1}\hat{g}]^{-1}$ is the alternate companion matrix $\hat{F}$ for f, where

$$\hat{g} = \begin{bmatrix} 0 \\ 0 \\ \vdots \\ 1 \end{bmatrix}$$

5. The system (F,g) is coefficient assignable.
6. The system (F,g) is pole assignable.

Moreover, if (F,g) is reachable and if $X^n + a_{n-1}X^{n-1} + \cdots + a_1X + a_0$ is the characteristic polynomial for F, then given a monic polynomial $h = X^n + b_{n-1}X^{n-1} + \cdots + b_1X + b_0 \in R[X]$, h is the characteristic polynomial of $F - gK$ where

$$K = [b_0 - a_0, \ldots, b_{n-1} - a_{n-1}][\hat{g}, \hat{F}\hat{g}, \ldots, (\hat{F})^{n-1}\hat{g}] \times [g, Fg, \ldots, F^{n-1}g]^{-1}$$

Proof. (1) $\Longleftrightarrow$ (2): This is clear from the definitions.

Before proving that condition (2) implies condition (3), we record the following. If g is a cyclic vector for F, then $B = \{g, Fg, \ldots, F^{n-1}g\}$ is a basis for R^n, and moreover, $[g, Fg, \ldots, F^{n-1}g]$ is the change of basis matrix from the standard basis of R^n to the basis B. In particular,

$$[g, Fg, \ldots, F^{n-1}g]^{-1}F[g, Fg, \ldots, F^{n-1}g]$$

is the matrix of F relative to the basis B. But this is easily seen to be the companion matrix for the characteristic polynomial of F. The only possible difficulty in seeing this occurs in the last column. If $X^n + a_{n-1}X^{n-1} + \cdots + a_1X + a_0$ is the characteristic polynomial for F, then

$$F^n = -a_{n-1}F^{n-1} - \cdots - a_1F - a_0$$

and so,

$$F^ng = F(F^{n-1}g) = -a_{n-1}F^{n-1}g - \cdots - a_1Fg - a_0g$$

From this remark, we see that (2) $\Longrightarrow$ (3).

(3) $\Longrightarrow$ (4): As for this implication, we make use of the above remark as well as the readily verifiable fact that the vector

$$\hat{g} = \begin{bmatrix} 0 \\ 0 \\ \vdots \\ 1 \end{bmatrix}$$

is a cyclic vector for the alternate companion matrix for $X^n + a_{n-1}X^{n-1} + \cdots + a_1X + a_0$. Thus,

$$[\hat{g},\hat{F}\hat{g},\ldots,(\hat{F})^{n-1}\hat{g}]^{-1}\hat{F}[\hat{g},\hat{F}\hat{g},\ldots,(\hat{F})^{n-1}\hat{g}] = \bar{F}$$

from which the result follows.

(5) ⟹ (6): This is obvious.

(6) ⟹ (1): This follows from Theorem 3.1.

Hence, the only remaining implication and the only one with substance is

(4) ⟹ (5): Set

$$P = [g,Fg,\ldots,F^{n-1}g] \text{ and } Q = [\hat{g},\hat{F}\hat{g},\ldots,(\hat{F})^{n-1}\hat{g}]$$

Then by (4), $QP^{-1}FPQ^{-1} = \hat{F}$.

Given a monic polynomial $h(X) = X^n + b_{n-1}X^{n-1} + \cdots + b_1X + b_0$, we would like to find a $1 \times n$ matrix K so that

$$\begin{aligned} h(X) &= \det(XI - F + gK) \\ &= \det(XI - (PQ^{-1})^{-1}F(PQ^{-1}) + (PQ^{-1})^{-1}gK(PQ^{-1})) \\ &= \det(XI - \hat{F} + \hat{g}K(PQ^{-1})) \end{aligned}$$

Thus, we wish to choose the matrix K so that

$$-\hat{F} + \hat{g}K(PQ^{-1}) = \begin{bmatrix} 0 & -1 & 0 & . & . & . & 0 \\ 0 & 0 & -1 & . & . & . & 0 \\ \vdots & & & & & & \\ a_0 & a_1 & a_2 & . & . & . & a_{n-1} \end{bmatrix} + \hat{g}K(PQ^{-1})$$

$$= \begin{bmatrix} 0 & -1 & 0 & . & . & . & 0 \\ 0 & 0 & -1 & . & . & . & 0 \\ \vdots & & & & & & \\ b_0 & b_1 & b_2 & . & . & . & b_{n-1} \end{bmatrix}$$

Consequently,

$$\hat{g}K(PQ^{-1}) = \begin{bmatrix} 0 & . & . & . & 0 \\ 0 & . & . & . & 0 \\ \vdots & & & & \\ b_0 - a_0 & . & . & . & b_{n-1} - a_{n-1} \end{bmatrix}$$

from which it follows that

$$\hat{g}K = \begin{bmatrix} 0 & \cdots & 0 \\ 0 & \cdots & 0 \\ \vdots & & \\ b_0 - a_0 & \cdots & b_{n-1} - a_{n-1} \end{bmatrix} (QP^{-1})$$

Writing $K = [k_1,\ldots,k_n]$, we see that

$$\hat{g}K = \begin{bmatrix} 0 & 0 & \cdot & \cdot & \cdot & 0 \\ 0 & 0 & \cdot & \cdot & \cdot & 0 \\ \vdots & & & & & \\ k_1 & k_2 & \cdot & \cdot & \cdot & k_n \end{bmatrix}$$

$$= \begin{bmatrix} 0 & \cdots & 0 \\ 0 & \cdots & 0 \\ \vdots & & \\ b_0 - a_0 & \cdots & b_{n-1} - a_{n-1} \end{bmatrix} (QP^{-1})$$

and hence that

$$K = [b_0 - a_0,\ldots,b_{n-1} - a_{n-1}](QP^{-1})$$

This proves not only that condition (4) implies condition (5), but also the moreover assertion.

So, the single-input case admits a complete treatment. Not only that, the proof we gave was constructive and demonstrated how to obtain a "feedback matrix" K.

For some rings, we are able to reduce the multi-input case to the single-input case. This is really just a corollary to the last result.

THEOREM 3.3. Given a system (F,G) over a ring R, suppose that there is an $m \times n$ matrix L such that $F - GL$ has cyclic vector $g = Gu$ for some vector $u \in R^m$. Then (F,G) is coefficient assignable. Therefore, if R is such that given any reachable system (F,G) there exists a vector $u \in R^m$ and an $m \times n$ matrix L such that $(F - GL, Gu)$ is reachable, then R has the coefficient assignability property.

Proof. By assumption, the system $(F - GL, Gu)$ is reachable with Gu as cyclic vector for $F - GL$. Hence, if a monic polynomial f is given, we can find a $1 \times n$ matrix K such that the characteristic polynomial of $(F - GL - GuK)$ is f. But, $F - GL - GuK = F - G(L - uK)$ and so (F,G) is coefficient assignable.

We shall see presently that the above theorem is useful for a large class of rings—for example, for fields and semi-quasi-local rings (= rings with only a finite number of maximal ideals). However, as we shall also see, it does not apply to the ring of integers nor to the polynomial ring in one variable over the field of real numbers. On the other hand, we shall prove that we can assign poles over these rings.

So, we would like to know for which rings R we can reduce to the single-input case. Let us say that a ring R has the *feedback cyclization property* if, whenever (F,G) is a reachable system over R, there exists a matrix K and a vector u such that Gu is a cyclic vector for the matrix $F - GK$. We proceed to our main theorem by a succession of lemmas.

LEMMA 3.4. A field has the feedback cyclization property.

Proof. Let R be a field. Assume that (F,G) is reachable so that the column space of the matrix $[G, FG, \ldots, F^{n-1}G]$ is R^n. Let g_1 be any nonzero column of G and consider $[g_1, Fg_1, \ldots, F^{n_1-1}g_1]$, where n_1 is the first integer ℓ such that $F^\ell g_1$ belongs to the span of the set $\{g_1, \ldots, F^{\ell-1}g_1\}$. Let g_2 be a column of G such that g_2 does not belong to the span of the set $g_1, \ldots, F^{n_1-1}g_1$ and let n_2 be the first integer ℓ such that $F^\ell g_2$ belongs to the span of the set $\{g_1, Fg_1, \ldots, F^{n_1-1}g_1, g_2, Fg_2, \ldots, F^{\ell-1}g_2\}$. Continuing in this fashion we will, <u>since R is a field</u>, eventually obtain a basis $\{g_1, Fg_1, \ldots, F^{n_1-1}g_1, \ldots, g_r, \ldots, F^{n_r-1}g_r\}$ for R^n. Next, we define $K : R^n \longrightarrow R^m$ as follows:

$$K(F^i g_j) = 0 \quad \text{if } i < n_j - 1$$

and, letting $c(g_i)$ denote the number of the column of G that g_i actually is,

$$K(F^{n_j-1} g_j) = \varepsilon_{c(g_{j+1})} \text{ if } j < r \text{ and } K(F^{n_r-1} g_r) = 0$$

(Here, ε_k denotes the kth elementary basis vector.)

We claim that $(F + GK, G\varepsilon_{c(g_1)})$ is reachable. First of all, note that $G\varepsilon_{c(g_1)} = g_1$. To verify the claim is a tedious calculation, but we illustrate what is involved:

$$(F + GK)g_1 = \begin{cases} Fg_1 & \text{if } n_1 \neq 1 \\ Fg_1 + g_2 & \text{if } n_1 = 1 \end{cases}$$

But $n_1 = 1$ implies that Fg_1 is a multiple of g_1.

$$(F + GK)^2 g_1 = (F + GK)Fg_1 = F^2 g_1 \text{ unless } n_1 = 2$$

$$(F + GK)^3 g_1 = F^3 g_1 \text{ unless } n_1 = 3$$

$$\vdots$$

$$(F + GK)^{n_1} g_1 = (F + GK)F^{n_1-1} g_1 = F^{n_1} g_1 + g_2$$

But, $F^{n_1} g_1$ belongs to the span of the set $\{g_1, \ldots, F^{n_1-1} g_1\}$. This should point the reader in the right direction and is as far as we shall go.

LEMMA 3.5. If $\{R_\alpha\}$ is a collection of rings having the feedback cyclization property, then their direct product also has the feedback cyclization property.

Proof. Let (F,G) be a reachable system over $R = \prod_\alpha R_\alpha$. If F_α is the matrix whose ijth entry is the αth coordinate of the ijth entry of F and G_α is defined similarly, then it is easily seen that (F_α, G_α) is a reachable system over R_α. Let K_α be a matrix and u_α a vector such that $G_\alpha u_\alpha$ is a cyclic vector for

$F_\alpha - G_\alpha K_\alpha$. Let $u = (u_\alpha)$ and let K be the matrix whose ijth coordinate is $((K_\alpha)_{ij})$. We claim that Gu is a cyclic vector for F - GK. To see this, note that the αth coordinate of

$$d = \det([Gu, (F - GK)Gu, \ldots, (F - GK)^{n-1}Gu])$$

is

$$\det([G_\alpha u_\alpha, (F_\alpha - G_\alpha K_\alpha)G_\alpha u_\alpha, \ldots, (F_\alpha - G_\alpha K_\alpha)^{n-1}G_\alpha u_\alpha])$$

which is a unit of the ring R_α. Hence, d is a unit in R, and Gu is a cyclic vector for F - GK.

LEMMA 3.6. Let R be a ring with Jacobson radical J. Then R/J has the feedback cyclization property if and only if R has the feedback cyclization property.

Proof. Suppose that R/J has the feedback cyclization property. Recall that J is nothing more than the intersection of the collection of all maximal ideals of R. If (F,G) is a reachable system over R, then the columns of the matrix $[G, FG, \ldots, F^{n-1}G]$ span R^n and so the columns of $[\bar{G}, \overline{FG}, \ldots, (\bar{F})^{n-1}\bar{G}]$ span $(\bar{R})^n = (R/J)^n$. Here, we are letting "–" denote images in the ring R/J. Since we are assuming that R/J has the feedback cyclization property, there exists a vector $u \in R^m$ and an $m \times n$ matrix K over R such that $\overline{Gu}$ is a cyclic vector over R/J for $\bar{F} - \overline{GK}$. This means precisely that

$$\det([\overline{Gu}, (\bar{F} - \overline{GK})\overline{Gu}, \ldots, (\bar{F} - \overline{GK})^{n-1}\overline{Gu}]) = \bar{b}$$

is a unit of R/J—that is,

$$\det([Gu, (F - GK)Gu, \ldots, (F - GK)^{n-1}Gu]) = b$$

does not belong to any maximal ideal of R. Thus, b is a unit of R and Gu is a cyclic vector for F - GK.

Conversely, suppose that R has the feedback cyclization property and let $(\bar{F}, \bar{G})$ be a reachable system over R/J. Then the span of the columns of $[\bar{G}, \overline{FG}, \ldots, \bar{F}^{n-1}\bar{G}]$ is $(R/J)^n$. Let F and G be matrices over R that give $\overline{F}$ and $\overline{G}$ upon passage to R/J. Then, for example,

$$\begin{bmatrix} 1 \\ 0 \\ 0 \\ \vdots \\ 0 \end{bmatrix} = C_1 + (\text{linear combination of the columns of } [G,\ldots,F^{n-1}G])$$

where $C_1 \in J^n$. Since this can be done for any of the standard basis vectors for R^n, there exist columns $C_1,\ldots,C_n \in J^n$ such that the system $(F,[G,C_1,\ldots,C_n])$ is reachable over R. Since R has the feedback cyclization property, there exists a matrix

$$K = \begin{bmatrix} K_1 \\ K_2 \end{bmatrix}$$

and a vector

$$u = \begin{bmatrix} u_1 \\ u_2 \end{bmatrix} \in R^{m+n}$$

such that $[G,C_1,\ldots,C_n]u$ is a cyclic vector for $F - [G,C_1,\ldots,C_n]K$. (Here, we are assuming that K_1 is $m \times n$ and K_2 is $n \times n$.) This means precisely that

$$\det([[G,C_1,\ldots,C_n]u,(F - [G,C_1,\ldots,C_n]K)[G,C_1,\ldots,C_n]u,\ldots]) = \varepsilon$$

a unit of R. Expanding this determinant, we see that

$$\varepsilon = \det([Gu_1,(F - GK_1)Gu_1,\ldots,(F - GK_1)^{n-1}Gu_1]) + t$$

where $t \in J$. This implies that $\overline{G}\overline{u}_1$ is a cyclic vector for $\overline{F} - \overline{G}\overline{K}_1$.

REMARK. Note that the second part of the proof actually works for any ideal of R, not merely for the Jacobson radical. Similar arguments can be used to prove that the pole assignability property and the coefficient assignability property are preserved under passage to homomorphic images.

We are now ready for the principal positive result of this section.

THEOREM 3.7. Let R be a ring having only finitely many maximal ideals. Then R has the feedback cyclization property and hence the coefficient assignability property.

Proof. Let $M_1, \ldots, M_r$ be the maximal ideals of R. Then $J = M_1 \cap \ldots \cap M_r$ is the Jacobson radical of R, and it is a fact from commutative algebra that R/J is a direct sum of fields [38, p. 109]. By Lemmas 3.4, 3.5, and 3.6, R has the feedback cyclization property. The final assertion follows from Theorem 3.3.

Theorem 3.7 gives many examples of rings having the feedback cyclization property and hence of rings having the pole assignability property. Indeed, since any quasi-local ring (that is, a ring with a unique maximal ideal) has the pole assignability property, the problem of determining the rings with the pole assignability property is reminiscent of the problem of determining those rings for which all finitely generated projective modules are free [51, p. 138]. We shall encounter additional connections later, but now we turn to the task of showing that neither of the rings Z nor $\mathbf{R}[X]$ has the feedback cyclization property.

THEOREM 3.8. Let R be a commutative ring. If R has the feedback cyclization property, then given $a,b \in R$ with $(a,b) = R$, there exist elements $c,d \in R$, d being a unit, such that $bc^2 \equiv d \pmod{a}$.

Proof. Let

$$F = \begin{bmatrix} 0 & 0 \\ b & 0 \end{bmatrix} \quad \text{and} \quad G = \begin{bmatrix} 1 & 0 \\ 0 & a \end{bmatrix}$$

Then the column module of the matrix $[G, FG]$ is R^2 since $(a,b) = R$. Since R has the feedback cyclization property, there exists a 2×2 matrix $K = [k_{ij}]$ over R and a vector

$$u = \begin{bmatrix} c \\ e \end{bmatrix} \in R^2$$

such that Gu is a cyclic vector for $F - GK$. Thus, $\det([Gu, (F - GK)Gu]) = d$ is a unit of R. Expanding the determinant, we see that

$$d = bc^2 + a(-c^2k_{21} - acek_{22} + eck_{11} + ae^2k_{12})$$

and hence that $bc^2 \equiv d \pmod a$.

Using this clever little theorem we can see why neither $\mathbf{R}[X]$ nor Z has the feedback cyclization property.

In $\mathbf{R}[X]$, consider $a = X^2 - 1$ and $b = X$. Clearly, a and b are relatively prime, and if $\mathbf{R}[X]$ had the feedback cyclization property, then there would exist polynomials h(x) and $g(x) \in \mathbf{R}[X]$ and a nonzero real number d such that

$$h(X) \cdot (X^2 - 1) = ([g(X)]^2 \cdot X) - d$$

Since this must hold for all real numbers, specialize to $X = 1$ and $X = -1$. Then

$$0 = [g(1)]^2(1) - d = [g(-1)]^2(-1) - d$$

Thus, the nonzero real number d is both positive and negative, which is absurd.

In the case of the ring Z of integers, consider $a = 5$ and $b = 2$. If Z had the feedback cyclization property, then there would have to be an integer c such that $2c^2 \equiv \pm 1 \pmod 5$. But working modulo 5, we have only to consider 0,1,2,3, and 4 to see that no such integer c can exist.

REMARK. Let **C** denote the field of complex numbers. It is an open problem to determine whether or not **C**[X] has the feedback cyclization property.

3.2. PROJECTIVE MODULES, BÉZOUT DOMAINS, AND THE POLE ASSIGNABILITY PROPERTY

In the previous section we saw that the rings Z and **R**[X] do not have the feedback cyclization property. However, each of these rings is a principal ideal domain—that is, an integral domain each of whose ideals is principal and henceforth denoted PID—and PID's do have the pole assignability property. This will follow from our work in this section.

The principal result of the section is Theorem 3.10, a corollary of which shows that PID's have the pole assignability property. In the process we shall be led to the investigation of the following question.

If D is a Bézout domain, does D have the pole assignability property?

Another corollary of Theorem 3.10 characterizes Bézout domains with the pole assignability property. Even with the corollary, we shall leave the question itself unresolved. We shall, however, introduce the notion of an elementary divisor ring and this will enable us to give several instances in which the question has an affirmative answer.

Preliminary to proving Theorem 3.10, we shall prove a result relating rings with the pole assignability property to rings having the property that certain projective modules over them are free. Before doing so, we must acquire the necessary notation and background. Our first task is to discuss bases for free modules. In a sense we shall be discussing why Theorem 3.7 isn't valid over arbitrary rings.

Let R be a commutative ring with

$$v = \begin{bmatrix} a_1 \\ a_2 \\ \vdots \\ a_n \end{bmatrix}$$

an element of R^n. Then v is part of a basis for R^n if and only if there exist vectors $v_2, \ldots, v_n \in R^n$ such that the matrix $[v, v_2, \ldots, v_n]$ is invertible if and only if $\det([v, v_2, \ldots, v_n])$ is a unit of R. If v as above <u>is</u> part of a basis, then by computing the determinant using the cofactor expansion on the column v, we see that

$$\det([v, v_2, \ldots, v_n]) = r_1 a_1 + \cdots + r_n a_n$$

a unit of R, and hence that $(a_1, \ldots, a_n) = R$. Thusly motivated, we define a vector

$$w = \begin{bmatrix} b_1 \\ b_2 \\ \vdots \\ b_n \end{bmatrix}$$

to be *unimodular* if $(b_1, \ldots, b_n) = R$. We have just shown that if w is part of a basis for R^n, then w is unimodular.

Conversely, given a unimodular vector

$$w = \begin{bmatrix} b_1 \\ b_2 \\ \vdots \\ b_n \end{bmatrix}$$

must w be part of a basis for R^n? Suppose the notation is $\sum_{i=1}^{n} r_i b_i = 1$. Consider the map $\tau : R^n \longrightarrow R$ defined by

$$\tau \begin{bmatrix} s_1 \\ s_2 \\ \vdots \\ s_n \end{bmatrix} = \sum_{i=1}^{n} s_i r_i$$

We see easily that τ is an R-homomorphism and that, since $\sum_{i=1}^{n} r_i b_i = 1$, τ is surjective. The projectivity of the R-module R implies that the exact sequence $R^n \xrightarrow{\tau} R \longrightarrow 0$ splits and that $R^n = \text{Ker}(\tau) \oplus \langle w \rangle$. If $\text{Ker}(\tau)$ is a free R-module, then $\text{Ker}(\tau) \cong R^{n-1}$, and thus, we can find vectors $w_2, \ldots, w_n \in R^n$ such that $\{w, w_2, \ldots, w_n\}$ is a basis for R^n. In general, $\text{Ker}(\tau)$ need not be free, but in certain instances it is. For example, since $\text{Ker}(\tau)$ is a direct summand of the free module R^n, $\text{Ker}(\tau)$ is a projective R-module. Thus, if R has the property that each finitely generated projective R-module is free, then $\text{Ker}(\tau)$ will be free. In this case, w is part of a basis for R^n. Of particular importance to us is the fact that Bézout domains and polynomial rings (in several variables) over a field have this property. (See Theorem 1.20 and [51, Theorem 4.59].)

If R is a ring and if B is a matrix over R, we shall say that B is a *good matrix* if there exists a matrix A such that the system (A,B) is reachable. If C is a matrix over R, then following [23] we shall call C a (*)-*matrix* if the ideal of R generated by the entries of C equals R and if all 2×2 minors of C are zero.

Finally, if P is a projective module over a ring R, then P is said to have *rank one* if for each maximal ideal M of R, the (R/M)-vector space P/MP is one-dimensional.

THEOREM 3.9. Suppose that a ring R has the pole assignability property. If G is a good matrix over R, there exists a matrix

V such that GV is a (*)-matrix. Moreover, if rank one projective R-modules are free, then G has a unimodular vector in its image.

Proof. Let G be an $n \times m$ good matrix over R. We must find a matrix V such that GV is a (*)-matrix. Now, since G is good and R has the pole assignability property, there exist an $n \times n$ matrix F and an $m \times n$ matrix K such that (F,G) is reachable and $F + GK$ has characteristic polynomial $(X - 1)^n$. Thus, $F + GK$ is invertible. Since $(F + GK,G)$ is also reachable, we may assume from the start that F is invertible.

If I is the $n \times n$ identity matrix then $(F + I,G)$ is reachable. By the pole assignability property there is an $m \times n$ matrix K′ for which $A = F + I + GK'$ has characteristic polynomial $X^{n-1}(X - 1)$. By the Cayley-Hamilton theorem, $A^{n-1}(A - I) = 0$ and it follows that A^{n-1} is an idempotent matrix. We claim, moreover, that A^{n-1} is a (*)-matrix.

First we show that the ideal generated by the entries of A^{n-1} is R. If not, there exists a maximal ideal M of R containing the entries of A^{n-1}. Using "‾" to denote reduction modulo M, we have that $\bar{A}^{n-1} = \bar{0}$. Thus, the minimal polynomial for $\bar{A}$ over the field $\bar{R}$ has no nonzero roots. On the other hand, the characteristic polynomial for $\bar{A}$ is $X^{n-1}(X - \bar{1})$, which has the nonzero root $\bar{1}$. This is a contradiction since over a field the roots of the characteristic polynomial and minimal polynomial must agree.

Next we show that all 2×2 minors of A^{n-1} are zero. We claim that the ideal J of R generated by the 2×2 minors of A^{n-1} is contained in every prime ideal of R. It will follow that each 2×2 minor of A^{n-1} is nilpotent [2, p. 5]. Let P be a prime ideal of R, L the quotient field of R/P and "‾" denote reduction modulo P. Since the characteristic roots of $\bar{A}$ are in L, there exists an invertible matrix Q over L with

$$Q\bar{A}Q^{-1} = \left[\begin{array}{cccc|c} 0 & 1 & & 0 & 0 \\ & \ddots & \ddots & & \vdots \\ & & \ddots & 1 & \vdots \\ & & & 0 & 0 \\ \hline 0 & \ldots & & 0 & 1 \end{array}\right]$$

a Jordan canonical form for $\bar{A}$ over L. Thus,

$$Q\bar{A}^{n-1}Q^{-1} = \left[\begin{array}{ccc|c} & & & 0 \\ & 0 & & \vdots \\ & & & 0 \\ \hline 0 & \ldots & 0 & 1 \end{array}\right]$$

and the 2×2 minors of $Q\bar{A}^{n-1}Q^{-1}$ are zero. Hence, the 2×2 minors of $\bar{A}^{n-1}$ vanish (since over the field L this is equivalent to rank $\bar{A}^{n-1} = 1$). Thus $J \subseteq P$ as claimed.

Now, since J is finitely generated by nilpotent elements, there exists a positive integer N such that $J^N = 0$. On the other hand, since A^{n-1} is an idempotent matrix, the Binet-Cauchy formula ([42, Theorem I.5]) implies that J is an idempotent ideal. Hence, $J = J^2 = \ldots = J^N = 0$, and all 2×2 minors of A^{n-1} are zero. Therefore, A^{n-1} is a (*)-matrix.

To finish, we have

$$0 = (A - I)A^{n-1} = (F + GK')A^{n-1}$$

so $GK'A^{n-1} = -FA^{n-1}$. Since F is invertible, it follows that $-FA^{n-1}$ is also a (*)-matrix (use the Binet-Cauchy formula to see that the 2×2 minors of $-FA^{n-1}$ are zero). Setting $V = K'A^{n-1}$, we have that GV is a (*)-matrix.

Now, suppose that rank one projective R-modules are free. Since the matrix A^{n-1} is idempotent, its column module C is a projective summand of R^n, and since the 2×2 minors of A^{n-1} vanish, C has rank one. Thus, C is free of rank one (exercise) and so $C = Rv$ for some element $v \in R^n$. Moreover, since the ideal generated by the entries of A^{n-1} is R, v must be unimodular. Thus, there is a unimodular vector v in the image of A^{n-1}. Finally, since F is invertible and $GK'A^{n-1} = -FA^{n-1}$, it follows that G has a unimodular vector in its image. The proof is now complete.

REMARK. In [25], M. Hautus and E. Sontag prove a form of the converse. Namely, suppose that a commutative ring R has the following property:

> Given a matrix G over R such that the ideal of R generated by the entries of G is R, there exists a matrix V such that GV is a (*)-matrix.

Then R has the pole assignability property.

Hautus and Sontag use this to prove that a Dedekind domain has the pole assignability property. We shall not give the proofs, but do wish to point out that as there exist Dedekind domains over which not all rank one projective modules are free, the latter property need not be possessed by rings having the pole assignability property.

To prove our main result, we need some more terminology. If R is a commutative ring, then following [51] we say that R has the *unimodular column property* if each unimodular vector over R is part of a basis—for example, if $u \in R^t$ is unimodular, then u is part of a basis for R^t. It is know that if R is such that finitely generated projective R-modules are free, then R has the unimodular column property. Indeed, R has the unimodular column property if and only if each stably free (projective) R-module is free. (An R-module E is called *stably free* if and only if there exist positive integers m and n such that $E \oplus R^m \cong R^n$.) See [42, Theorem IV.41].

There is one additional important definition. Let (F,G) be a system over a ring R. Call a system $(\tilde{F},\tilde{G})$ *systems equivalent* to (F,G) if it is obtained from (F,G) by one of the following transformations:

1. $F \longmapsto \tilde{F} = AFA^{-1}$, $G \longmapsto \tilde{G} = AG$ for some invertible matrix A. This type of transformation is a consequence of a change of basis in R^n, the state module.

2. $F \longmapsto \tilde{F} = F$, $G \longmapsto \tilde{G} = GB$ for some invertible matrix B. This type of transformation is a consequence of a change of basis in R^m, the input module.

3. $F \longmapsto \tilde{F} = F + GK$, $G \longmapsto \tilde{G} = G$ for any matrix K of suitable size. This type of transformation is a consequence of state feedback.

It is clear that systems equivalence is an equivalence relation. Moreover, if (F,G) is systems equivalent to $(\tilde{F},\tilde{G})$, then (F,G) is reachable (resp. pole assignable, coefficient assignable) if and only if $(\tilde{F},\tilde{G})$ is reachable (resp. pole assignable, coefficient assignable).

THEOREM 3.10. Let R be a commutative ring and suppose that rank one projective R-modules are free. The following are equivalent.

1. R has the pole assignability property.

2. Each good matrix over R has a unimodular vector in its image.

3. Each reachable system over R is systems equivalent to one of the form

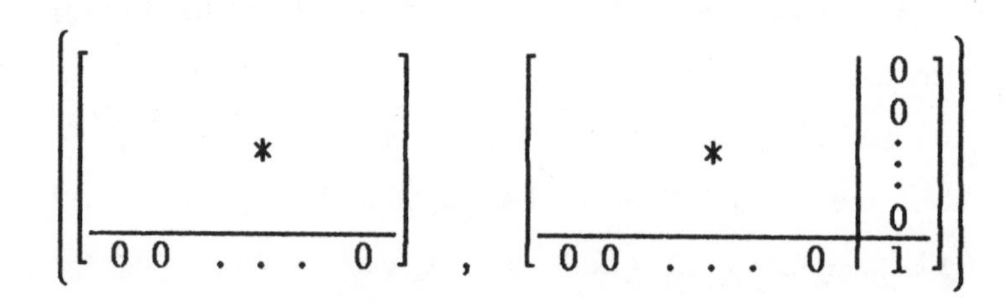

Proof. (1) ⟹ (2): This follows from Theorem 3.9.

(2) ⟹ (3): We first prove that stably free R-modules are free. As noted prior to the theorem, it will follow that

R has the unimodular column property. So let P be a stably free R-module. By [42, Theorem IV.44], there exist positive integers m and n such that P^m is isomorphic to R^n as R-modules. We proceed to find a reachable system (F,G) such that the column module of G is isomorphic to P.

Now, without loss of generality, we may assume that $R^n = P \oplus \cdots \oplus P$, m times. Let $g : R^n \longrightarrow R^n$ be projection onto the first P-factor and let $f : R^n \longrightarrow R^n$ be defined as $f(p_1,\ldots,p_m) = (p_m,p_1,\ldots,p_{m-1})$. Using im(g) to denote the image of g, we clearly have

$$\langle \mathrm{im}(g), f(\mathrm{im}(g)), \ldots, f^{n-1}(\mathrm{im}(g))\rangle = R^n$$

Thus, if F and G are any matrices representing f and g respectively, it follows that (F,G) is a reachable system with the column module of G isomorphic to P. By hypothesis (2), the column module of G contains a unimodular vector and consequently P contains a free summand of rank one. By induction on the rank of P, it follows that P is free. (See Exercises at the end of the chapter.)

Now, suppose that (F,G) is an arbitrary reachable system over R. By hypothesis (2), there exist unimodular vectors $u \in R^m$ and $v \in R^n$ such that $Gu = v$. Since we have shown that R has the unimodular column property, there exist elements $u_1,\ldots,u_{m-1} \in R^m$ and $v_1,\ldots,v_{n-1} \in R^n$ such that $\{u_1,\ldots,u_{m-1},u\}$ is a basis for R^m and $\{v_1,\ldots,v_{n-1},v\}$ is a basis for R^n. Let $U = [u_1,\ldots,u_{m-1},u]$ and $V = [v_1,\ldots,v_{n-1},v]$. Then (F,G) is systems equivalent to $(V^{-1}FV, V^{-1}GU)$, where $V^{-1}GU$ has the form

$$\left[\begin{array}{c|c} & 0 \\ * & \vdots \\ & 0 \\ & 1 \end{array}\right]$$

Right multiplication by an invertible matrix (column operations) shows that this latter system is equivalent to $(\tilde{F},\tilde{G})$, where $\tilde{F} = V^{-1}FV$ and

$$\tilde{G} = \left[\begin{array}{c|c} * & \begin{matrix}0\\ \vdots \\ 0\end{matrix} \\ \hline 0 \ \cdots \ 0 & 1 \end{array}\right]$$

Now let

$$K = \left[\begin{array}{c} 0 \\ \hline -a_1 \ \cdots \ -a_n \end{array}\right]$$

where $[a_1,\ldots,a_n]$ is the last row of $\tilde{F}$. Then the system $(\tilde{F} + \tilde{G}K, \tilde{G})$ has the desired form. This proves (3).

(3) $\Longrightarrow$ (1): We use induction on n, the dimension of the reachable system (F,G). If n = 1, then (F,G) has the form $([a],[b_1,\ldots,b_m])$ for elements $a,b_1,\ldots,b_m \in R$. Since (F,G) is reachable, there exist elements $t_1,\ldots,t_m \in R$ such that $t_1b_1 + \cdots + t_mb_m = 1$. If $r \in R$, set

$$K = (a - r)\begin{bmatrix} t_1 \\ \vdots \\ t_m \end{bmatrix}$$

Then $F - GK = [r]$ and so $\det(XI - (F - GK)) = X - r$.

If n > 1 we can, by hypothesis, assume that (F,G) has the form

$$\left(\left[\begin{array}{c|c} F_1 & G_1 \\ \hline 0 & 0 \end{array}\right], \left[\begin{array}{c|c} G_2 & \begin{matrix}0\\0\\ \vdots \\ 0\end{matrix} \\ \hline 0 \ \cdots \ 0 & 1 \end{array}\right]\right)$$

where F_1 is $(n - 1) \times (n - 1)$, G_2 is $(n - 1) \times (m - 1)$, and $G_1 \in R^{n-1}$. We claim that the $(n - 1)$-dimensional system $(F_1, [G_1, G_2])$ is reachable. This will follow from the next lemma.

LEMMA. Let (A,B) be a system over a ring R and suppose that (A,B) has the following form

$$(A,B) = \left(\begin{bmatrix} F_1 & G_1 \\ H_1 & A_1 \end{bmatrix}, \begin{bmatrix} G_2 & 0 \\ 0 & B_1 \end{bmatrix} \right)$$

with B_1 invertible. Then (A,B) is reachable if and only if $(F_1,[G_1,G_2])$ is reachable.

Proof of lemma. Write down the corresponding reachability matrices. These are

$$\Big[[G_1,G_2],\ [F_1G_1,F_1G_2],\ [F_1{}^2G_1,F_1{}^2G_2],\ldots\Big] \tag{i}$$

and

$$\left[\begin{bmatrix} G_2 & 0 \\ 0 & B_1 \end{bmatrix}, \begin{bmatrix} F_1G_2 & G_1B_1 \\ H_1G_2 & A_1B_1 \end{bmatrix}, \begin{bmatrix} F_1{}^2G_2 + G_1H_1G_2, & F_1G_1B_1 + G_1A_1B_1 \\ * & * \end{bmatrix}, \ldots\right] \tag{ii}$$

We now indicate why the lemma is true. The idea is to show that the columns of (i) and those of the first row of (ii) have the same span. Since B_1 is invertible, the result will follow. So, let S_1 denote the span of the columns of (i) and S_2 the span of the columns of the first row of (ii). Clearly, $S_1 \supseteq S_2$. On the other hand, each column of G_2 is in S_2, as is each column of G_1, since B_1 is invertible. Each column of F_1G_2 is in S_2 and, since each column of G_1 is, so is each column of $G_1A_1B_1$. But then each column of $F_1G_1B_1$, and hence of F_1G_1, is in S_2, etc.

Returning to the proof of the theorem, let $r_1,\ldots,r_{n-1}$, $r_n \in R$ be given. Since $(F_1,[G_1,G_2])$ is reachable of dimension $n-1$, there exists, by induction on n, an $m \times (n-1)$ matrix $\tilde{K}$ such that $F_1 + [G_1,G_2]\tilde{K}$ has characteristic polynomial $(X - r_1)\cdots(X - r_{n-1})$. Write

$$\tilde{K} = \begin{bmatrix} K_1 \\ K_2 \end{bmatrix}$$

where K_1 is $1 \times (n-1)$ and K_2 is $(m-1) \times (n-1)$. Then $F_1 + [G_1,G_2]\tilde{K} = F_1 + G_1K_1 + G_2K_2$. Note that (F,G) is systems equivalent to

$$\left(\begin{bmatrix} I & 0 \\ -K_1 & 1 \end{bmatrix}\begin{bmatrix} F_1 & G_1 \\ 0 & 0 \end{bmatrix}\begin{bmatrix} I & 0 \\ K_1 & 1 \end{bmatrix}, \begin{bmatrix} I & 0 \\ -K_1 & 1 \end{bmatrix}\begin{bmatrix} G_2 & 0 \\ 0 & 1 \end{bmatrix}\right)$$

$$= \left(\begin{bmatrix} F_1 + G_1K_1 & G_1 \\ * & * \end{bmatrix}, \begin{bmatrix} G_2 & 0 \\ -K_1G_2 & 1 \end{bmatrix}\right)$$

The latter system is equivalent to

$$(\tilde{F},\tilde{G}) = \left(\begin{bmatrix} F_1 + G_1K_1 & G_1 \\ * & * \end{bmatrix} + \begin{bmatrix} G_2 & 0 \\ -K_1G_2 & 1 \end{bmatrix}\begin{bmatrix} K_2 & 0 \\ 0 & 0 \end{bmatrix},\right.$$

$$\left.\begin{bmatrix} G_2 & 0 \\ -K_1G_2 & 1 \end{bmatrix}\begin{bmatrix} I & 0 \\ K_1G_2 & 1 \end{bmatrix}\right) = \left(\begin{bmatrix} F_1 + G_1K_1 + G_2K_2 & G_1 \\ * & * \end{bmatrix}, G\right)$$

Let K be the $m \times n$ matrix

$$\begin{bmatrix} & 0 & & \\ \hline -a_1 & \ldots & -a_{n-1} & r_n - a_n \end{bmatrix}$$

where $[a_1,\ldots,a_n]$ is the last row of $\tilde{F}$. Then

$$\tilde{F} + GK = \left[\begin{array}{c|c} F_1 + G_1K_1 + G_2K_2 & G_1 \\ \hline 0 \quad . \quad . \quad . \quad 0 & r_n \end{array}\right]$$

which has the desired characteristic polynomial $(X - r_1)\cdots(X - r_n)$. This completes the proof of Theorem 3.10.

Theorem 3.10 has several important corollaries. In the first two, we use the folklore fact that a rank one projective module is finitely generated.

THEOREM 3.11. Let R be a field or a PID and let $X_1,\ldots,X_n$ be indeterminates over R. Then $R[X_1,\ldots,X_n]$ has the pole assignability property if and only if each good matrix over $R[X_1,\ldots,X_n]$ has the pole assignability property if and only if each good matrix over $R[X_1,\ldots,X_n]$ has a unimodular vector in its image.

Proof. By the Quillen-Suslin theorem [51, Theorem 4.63] finitely generated projective modules are free over $R[X_1,\ldots,X_n]$.

THEOREM 3.12. Let D be a Bézout domain. Then D has the pole assignability property if and only if each good matrix over D has a unimodular vector in its image.

Proof. By Theorem 1.20, finitely generated projective modules over D are free.

A commutative ring R is said to be an *elementary divisor ring* if and only if given a matrix A over R, there exist invertible matrices P and Q over R such that

$$PAQ = \begin{bmatrix} r_1 & & 0 \\ & r_2 & \\ 0 & & \ddots \end{bmatrix}$$

That is, PAQ has zeros above and below the main diagonal.

THEOREM 3.13. An elementary divisor ring R has the pole assignability property.

Proof. By [64], rank one projective R-modules are free over such a ring. Consequently, to apply Theorem 3.10, we have to prove that every good matrix G over R has a unimodular vector in its image. Choose invertible matrices P and Q over R so that

$$PGQ = \begin{bmatrix} r_1 & & & 0 \\ & r_2 & & \\ & & \ddots & \\ 0 & & & r_n \end{bmatrix}$$

Since G is good, the ideal of R generated by the entries of G is R. It follows that $(r_1, r_2, \ldots, r_n) = R$. But then

$$GQ \begin{bmatrix} 1 \\ 1 \\ \vdots \\ 1 \end{bmatrix} = P^{-1} \begin{bmatrix} r_1 & & & 0 \\ & r_2 & & \\ & & \ddots & \\ 0 & & & r_n \end{bmatrix} \begin{bmatrix} 1 \\ 1 \\ \vdots \\ 1 \end{bmatrix}$$

is a unimodular vector in the image of G.

REMARK. It is classical that a principal ideal domain is an elementary divisor ring and therefore that any PID has the pole assignability property. This comment applies, in particular, to the rings Z or **R**[X]. On the other hand, we shall prove in the next section that neither Z[Y] nor **R**[X,Y] has the pole assignability property.

Whether or not every Bézout domain has the pole assignability property is still an open question. If, however, the following old conjecture is true, then the question has an affirmative answer.

CONJECTURE. Every Bézout domain is an elementary divisor domain.

At our present level of knowledge however, if we wish to show that a certain Bézout domain has the pole assignability property, we must prove that it's (almost) an elementary divisor domain.

We now prove two heretofore unpublished results of I. Kaplansky regarding when a Bézout domain is an elementary divisor domain. We gratefully acknowledge Professor Kaplansky's willingness to allow us to include them. The first of these says that cardinality of the family of maximal ideals is important while the second one says that, at least when the dimension of the Bézout domain is very small, the domain must be an elementary divisor domain.

THEOREM 3.14. Let D be a Bézout domain having only countably many maximal ideals. Then D is an elementary divisor domain.

Proof. It suffices to prove that each matrix of the form

$$A = \begin{bmatrix} a & b \\ c & d \end{bmatrix} \text{ with } (a,b,c,d) = 1$$

can be diagonalized (see [36]). Let M be a maximal ideal of D. We claim that A is equivalent to a matrix

$$\begin{bmatrix} a' & b' \\ c' & d' \end{bmatrix} \text{ where } a' \text{ divides } a,\ a' \notin M$$

Write $(a,c) = pa + qc$. Then

$$\begin{bmatrix} p & q \\ \frac{c}{(a,c)} & \frac{-a}{(a,c)} \end{bmatrix} \begin{bmatrix} a & b \\ c & d \end{bmatrix} = \begin{bmatrix} (a,c) & - \\ 0 & - \end{bmatrix}$$

Thus, A is equivalent to a matrix

$$\begin{bmatrix} a^* & b^* \\ 0 & d^* \end{bmatrix} \text{ where } a^* \text{ divides } a$$

If $a^* \notin M$, fine. Suppose that $b^* \notin M$ and that $(a^*,b^*) = xa^* + yb^*$. Then

$$\begin{bmatrix} a^* & b^* \\ 0 & d^* \end{bmatrix} \begin{bmatrix} x & - \\ y & - \end{bmatrix} = \begin{bmatrix} (a^*,b^*) & - \\ - & - \end{bmatrix}$$

where (a^*,b^*) divides a^* and hence (a^*,b^*) divides a, but (a^*,b^*) does not belong to M. If $b^* \in M$, but $d^* \notin M$, then

$$\begin{bmatrix} 1 & 1 \\ 0 & 1 \end{bmatrix} \begin{bmatrix} a^* & b^* \\ c^* & d^* \end{bmatrix} = \begin{bmatrix} a^* & b^* + d^* \\ 0 & d^* \end{bmatrix}$$

This returns us to the case just treated and the claim is justified.

Let $M_1, M_2, \ldots$ be the set of all maximal ideals of D. Do the above business to $M_1, M_2, \ldots$ in succession obtaining elements $a_1, a_2, \ldots \in D$ such that

$$(a) \subseteq (a_1) \subseteq (a_2) \subseteq \cdots$$

Then $\bigcup_{i=1}^{\infty}(a_i)$ is contained in no maximal ideal of D and so $\bigcup_{i=1}^{\infty}(a_i) = D$. It follows that some a_n is a unit of D and hence that A is equivalent to a matrix

$$\begin{bmatrix} a_n & - \\ - & - \end{bmatrix} \quad \text{where } a_n \text{ is a unit}$$

Such a matrix can easily be diagonalized and the proof is complete.

Let R be a commutative ring and let n be a positive integer. If there exists a chain

$$P_0 \subsetneq P_1 \subsetneq P_2 \subsetneq \cdots \subsetneq P_n$$

of prime ideals of R, but no longer such chain, then R is said to have *dimension* n. (If there is no bound to the lengths of chains of prime ideals of R, then R is said to be *infinite dimensional.*) For example, the ring Z has dimension one; the ring $Z[X_1, \ldots, X_n]$ has dimension n + 1; the ring $L[X_1, \ldots, X_n]$ has dimension n for L a field.

THEOREM 3.15. If D is a one-dimensional Bézout domain, then D is an elementary divisor domain. In fact, it is "adequate."

Proof. By [36, p. 473], we have to prove that if a and d are nonzero elements of D, then there exist elements $b, c \in D$ with $a = bc$, $(b,d) = 1$, and no nonunit factor of c is relatively prime to d. Consider the following sequence of <u>elements</u> of D:

$$a_1 = \frac{a}{(a,d)}, \quad a_2 = \frac{a_1}{(a_1,d)}, \quad a_3 = \frac{a_2}{(a_2,d)}, \ldots$$

We claim that for some integer n, $(a_n,d) = 1$. Otherwise, look at the following chain of <u>ideals</u> of D:

$$(a,d) \subseteq (a_1,d) \subseteq (a_2,d) \subseteq \cdots$$

The union is a proper ideal of D and so is contained in a maximal ideal M of D. Since D is a Bézout domain, it is a "Prüfer" domain. Consequently, D_M is a valuation domain—that is, an integral domain in which the ideals are linearly ordered. [See 22, p. 275.] Hence, for every i, $(a_i)D_M \supseteq (d)D_M$ or $(a_i)D_M \subseteq (d)D_M$. Thus, for every i, $(a_i,d)D_M = (a_i)D_M$ or $(a_i,d)D_M = (d)D_M$. If, for some i, $(a_i,d)D_M = (a_i)D_M$, then $(a_{i+1},d)D_M = D_M$, a contradiction. The alternative is that $(a_i,d)D_M = (d)D_M$ for each i. This implies that $a \in (d^i)D_M$ for each i and hence that $a = 0$, since D_M is a one-dimensional valuation domain [22, p. 187]. This justifies the claim.

Now, set $b = a_n$ and $c = a/b$. Then $(b,d) = 1$, $a = bc$, and we have only to see that no nonunit factor of c is relatively prime to d. To see this, observe that $c = (a,d)(a_1,d)\cdots(a_{n-1},d)$. If $c = rs$ for $r,s \in D$, r not a unit, let M be a maximal ideal of D containing r. Then $c \in M$ and so $(a_i,d) \subseteq M$ for some i. Thus, $d \in M$ and $r \in M$ from which it follows that $(r,d) \subseteq M$—that is, r and d are not relatively prime.

That D is an elementary divisor domain follows from [36, Theorem 5.3].

We now show that the ring R of real analytic functions, which was studied in Theorem 1.19, is an infinite-dimensional Bézout domain with uncountably many maximal ideals. Thus, neither Theorem 3.14 nor Theorem 3.15 is applicable, although R does turn out to be an elementary divisor domain.

Indeed, R is a Bézout domain by Theorem 1.19, and R has uncountably many maximal ideals since, for real t, $\{f \in R \mid f(t) = 0\}$ is a maximal ideal of R. To prove that R is infinite-dimensional, we need some facts about prime ideals in commutative rings, as well as some facts about real analytic functions. Let S be any commutative ring with unit, and let A be a proper ideal of S. Then A is contained in a maximal ideal of S, and maximal ideals are prime. Also, the intersection of a chain of prime ideals is easily seen to be a prime ideal. It follows from Zorn's lemma that there exists a prime ideal minimal among all prime ideals containing A.

LEMMA 3.16. If P is a minimal prime ideal containing an ideal A in a ring S, then if $f \in P$, there is an element $e \in S\backslash P$ and a natural number m such that $ef^m \in A$.

Proof. If not, then the set

$$M = \{ef^m \mid e \in S\backslash P,\ m = 1,2,3,\ldots\}$$

is a multiplicatively closed set that has empty intersection with A. Let Q be an ideal of S that contains A and is maximal with respect to having empty intersection with M. It is easily seen that Q is a prime ideal and obviously Q is contained in P. Since $f \in P$ and $f \notin Q$, Q is properly contained in P. This contradicts the minimality of P and completes the proof.

For x real and p = 1,2,3,..., let

$$Ep(x) = (1 - x) \exp\left(x + \frac{x^2}{2} + \cdots + \frac{x^p}{p}\right)$$

Let (x_n) be a sequence of real numbers such that $x_n \neq 0$ and $|x_n| \to \infty$ as $n \to \infty$. Then natural numbers p_n may be chosen so that the infinite product

$$f(x) = \prod_{n=1}^{\infty} Ep_n\left(\frac{x}{x_n}\right) \tag{*}$$

converges and defines a real analytic function that has a zero at each point x_n and has no other zeros; see [55, Theorem 15.9]. More precisely, if t occurs m times in the sequence (x_n), then f has a zero of order m at t. The function $x^m f(x)$ also has a zero of order m at 0.

If f is any real analytic function that is not identically zero, then the set of zeros of f is an at most countable set without a cluster point; see [15, 9.1.5]. If f has a zero of order r at t, then $f(x) = (x - t)^r g(x)$, where g is a real analytic function with $g(t) \neq 0$. Indeed, let $g(x) = f(x)/(x - t)^r$ if $x \neq t$ and let $g(t) = f^{(n)}(t)/n!$, where $f^{(n)}$ denotes the nth derivative of f. Then g is clearly analytic at each x, $x \neq t$, and by the proof of [15, 9.1.5], g is also analytic at t.

Let o(f)(t) denote the order of the zero of f at t, with o(f)(t) = 0 if $f(t) \neq 0$. If o(f)(t) = r, then

$$f(x) = (x - t)^r g(x)$$

where g is real analytic and $g(t) \neq 0$. We claim that if f_1 and f_2 are real analytic functions with $o(f_1)(t) \geq o(f_2)(t)$ for each t, then there is a real analytic function h with $f_1 = f_2 h$. Let $x_1, x_2, x_3, \ldots$ be the zeros of f_2; say x_n has order r_n. Then $f_2(x) = (x - x_n)^{r_n} g_n(x)$, where each g_n is real analytic and

$g_n(x_n) \neq 0$. Let $o(f_1)(x_n) = s_n \geq r_n$, with $f_1(x) = (x - x_n)^{s_n}h_n(x)$. Define h by $h(x) = f_1(x)/f_2(x)$ if $f_2(x) \neq 0$, $h(x_n) = 0$ if $s_n > r_n$, and $h(x_n) = h_n(x_n)/g_n(x_n)$ if $s_n = r_n$. Then h is real analytic and $f_1 = f_2h$.

Thusly prepared, we can now prove that R is infinite-dimensional.

THEOREM 3.17. The ring R of real analytic functions is infinite dimensional.

Proof. For i and r natural numbers let $f_{i,r}$ be the real analytic function, defined as in equation (*) above, with zeros at i, i + 1, i + 2,... and with $o(f_{i,r})(j) = j^r$ for each $j \geq i$. Then $o(f_{i,r+1}) \geq o(f_{i,r})$, so $f_{i,r+1} \in f_{i,r}R$. Let A_r be the ideal of R generated by the set $\{f_{i,r} \mid i = 1,2,3,\ldots\}$. Then $A_{r+1} \subseteq A_r$ for each r. Clearly, A_1 is a proper ideal of R. Let P_1 be a minimal prime ideal containing A_1, and let P_2 be a prime ideal that is minimal among the prime ideals containing A_2 and contained in P_1. We can inductively continue this process to obtain a sequence of prime ideals

$$P_1 \supseteq P_2 \supseteq P_3 \supseteq \cdots$$

such that P_r is a minimal prime ideal containing A_r. We claim that $P_r \neq P_{r+1}$ for every r. Suppose, for contradiction, that $P_r = P_{r+1}$. Since $f_{1,r} \in A_r \subseteq P_r = P_{r+1}$, it follows from Lemma 3.16 that there is an element $e \in R\backslash P_{r+1}$ and a natural number n such that $ef_{1,r}^n \in A_{r+1}$.

Hence,

$$ef_{1,r}^n = a_1f_{m_1,r+1} + a_2f_{m_2,r+1} + \cdots + a_qf_{m_q,r+1}$$

for some $m_1,m_2,\ldots,m_q$ and some $a_1,a_2,\ldots,a_q$ in R. If $m = \max\{m_1,\ldots,m_q\}$ and $i \geq m$, then

$$ni^r + o(e)(i) \geq i^{r+1}$$

For $i > p = \max\{n,m\}$,

$$o(e)(i) \geqslant i^{r+1} - ni^r \geqslant (n+1)i^r - ni^r = i^r$$

That is, $o(e) \geqslant o(f_{p+1,r})$ and hence $e \in f_{p+1,r}R \subseteq A_r$. This is a contradiction because $e \notin P_{r+1} \supseteq A_r$. Thus, the P_r's form an infinite chain of strictly decreasing prime ideals.

THEOREM 3.18. The ring R of real analytic functions is an elementary divisor domain. In fact, it is "adequate."

Proof. As in the proof of Theorem 3.15, we prove that if $a,c \in R$ with $a \neq 0$, then we can write $a = rs$ with $(r,c) = 1$ and $(u,c) \neq 1$ for any nonunit divisor u of s. Let (x_i) be the set of common zeros of a and c with n_i the order of a at x_i. Let s be a real analytic function with precisely these zeros and orders. Then $o(a) \geqslant o(s)$, so $a = rs$ for some $r \in R$. The zeros of (r,c) are exactly the common zeros, including orders, of r and c. If (r,c) were zero at some x, then x would have to be one of the x_i's and the order of a at x_i would be greater than n_i. Hence, (r,c) has no zeros and $(r,c) = 1$. If u is any nonunit divisor of s, then u must be zero at some x_i, so (u,c) is zero at x_i and $(u,c) \neq 1$. This completes the proof that R is adequate. That R is an elementary divisor domain follows from [36, Theorem 53.].

Combining Theorems 1.19, 3.17 and 3.18, we record the following theorem.

THEOREM 3.19. The ring of real analytic functions has the following properties:

1. It is completely integrally closed.

2. It is an elementary divisor domain and hence a Bézout domain.

3. It is infinite dimensional—that is, it has infinite Krull dimension.

We note that if J is any open interval of the real line, then the ring of real analytic functions on J is isomorphic to the ring of functions analytic on the entire real line. Indeed, since the inverse tangent function

$$f(z) = \int_0^z \frac{1}{1 + p^2}\, dp$$

is complex analytic on the strip $\{z \mid -1 < \text{Im} z < 1\}$, f is a real analytic function from the reals onto the interval $(-\pi/2, \pi/2)$. The mapping f_* defined by $f_*(\phi) = \phi \circ f$ is then easily seen to be a ring isomorphism from the ring of functions analytic on $(-\pi/2, \pi/2)$ to the ring of functions analytic on the entire real line. Likewise, the rings of functions analytic on any two open intervals are isomorphic.

We have now proven that the ring of real analytic functions on an open interval is an infinite dimensional elementary divisor domain and that the ring of real analytic functions on the unit circle is a Dedekind domain. (See Section 1.7.) Ring theoretically, these rings of functions are far apart, but both have the pole assignability property.

We end this section by noting that in [10] examples are given to show that neither Z nor $\mathbf{R}[Y]$ has the coefficient assignability property. In particular, a PID need not have the coefficient assignability property.

3.3 COUNTEREXAMPLES TO POLE ASSIGNABILITY

The converse to Theorem 3.1 is false—that is, there exist rings and reachable systems over them which are not pole assignable. We present such an example below. It demonstrates that even very nice rings need not have the pole assignability property and also illustrates the depth of the pole assignability question. There are several examples of this phenomenon and we shall prove the following theorem that delineates the pathology inherent in each of them. It will follow from Theorem 3.20 that a polynomial ring in one variable over the integers or in two variables over a field fails to have the pole assignability property.

THEOREM 3.20. Let D be a one-dimensional Bézout domain with Y an indeterminate. Suppose that there exists a prime polynomial $g \in D[Y]$ such that in the integral domain $(D[Y])/(g)$ there is an ideal I with the following properties:

1. I is generated by two elements,
2. I is not principal, but
3. I^2 is principal.

Then D[Y] does not have the pole assignability property.

Proof. For convenience, set $D_1 = (D[Y])/(g)$. By definition, since g is a prime polynomial, D_1 is an integral domain. Further, becaue I^2 is principal and hence invertible, I is invertible, which is equivalent to being projective as a D_1-module. (See Theorem 1.23.)

Suppose that $I = (x,z)$ and consider the natural surjection ϕ from $D_1 \oplus D_1 \longrightarrow I$ given by $\phi(1,0) = x$ and $\phi(0,1) = z$. Since I is projective, the sequence splits giving $D_1 \oplus D_1 \cong I \oplus \operatorname{Ker} \phi$. We claim that

$$\operatorname{Ker} \phi \cong I^{-1} = \{\alpha \in \text{quotient field of } D_1 \mid \alpha I \subseteq D_1\}$$

Since I is invertible, $II^{-1} = D_1$, say $1 = \alpha x + \beta z$ for $\alpha, \beta \in I^{-1}$. Define a map $\theta : \operatorname{Ker} \phi \longrightarrow I^{-1}$ by $\theta(a,b) = a\beta - b\alpha$. Then given $\gamma \in I^{-1}$,

$$\theta(\gamma z, -\gamma x) = \gamma z\beta + \gamma x\alpha = \gamma(z\beta + x\alpha) = \gamma$$

and

$$\phi(\gamma z, -\gamma x) = \gamma zx - \gamma xz = 0$$

Thus, $(\gamma z, -\gamma x) \in \operatorname{Ker} \phi$ and so θ is surjective. Moreover, if

$$(a,b) \in \operatorname{Ker} \phi \cap \operatorname{Ker} \theta$$

then $a\beta - b\alpha = 0 = ax + bz$. Then, for example,

$$0 = ax\beta - bx\alpha = ax\beta + bz\beta$$

It follows that $bz\beta + bx\alpha = 0$ and so $b(z\beta + x\alpha) = b = 0$. Similarly, $a = 0$. Thus, θ is an isomorphism and $D_1 \oplus D_1 \cong I \oplus I^{-1}$. But I^2 is principal, say $I^2 = (d)$. Now $(d^{-1})II = D_1$ and so $d^{-1}I = I^{-1}$, the unique inverse in the group of all invertible ideals of D_1. The map ψ: $I^{-1} \longrightarrow I$ given by $\psi(i/d) = i$ is a D_1-module isomorphism and so $I^{-1} \cong I$. From this it follows that $D_1 \oplus D_1 \cong I \oplus I$. Suppose that the notation is that ϕ is an isomorphism from $D_1 \oplus D_1$ onto $I \oplus I$. (This map ϕ is not to be confused with the ϕ defined above.) Consider the D_1-endomorphisms of $I \oplus I$ defined by

$$\tilde{F}(a,b) = (0,a), \ \tilde{P}(a,b) = (a,0)$$

These induce D_1-endomorphisms $\bar{F}$ and $\bar{P}$ of $D_1 \oplus D_1$ given by $\bar{F} = \phi^{-1} \circ \tilde{F} \circ \phi$ and $\bar{P} = \phi^{-1} \circ \tilde{P} \circ \phi$. We claim that $D_1 \oplus D_1 = \operatorname{Im}\bar{F} \oplus \operatorname{Im}\bar{P}$ and that $\operatorname{Im}\bar{F} \cong \operatorname{Im}\bar{P} \cong I$.

First, since ϕ is surjective, $\operatorname{Im}(\tilde{F} \circ \phi) = 0 \oplus I \cong I$ and so $\operatorname{Im}(\phi^{-1} \circ \tilde{F} \circ \phi) = \operatorname{Im}\bar{F} \cong I$ since ϕ^{-1} is an isomorphism. Similarly,

$$\operatorname{Im}(\phi^{-1} \circ \tilde{P} \circ \phi) = \operatorname{Im}\bar{P} = \phi^{-1}(I \oplus 0) \cong \phi^{-1}(I) \cong I$$

Next, if $\alpha = (i_1,i_2) \in I \oplus I$, then $\alpha = \tilde{F}(i_2,0) + \tilde{P}(i_1,0)$. Thus, given $\beta \in D_1 \oplus D_1$, we can find α_1, $\alpha_2 \in I \oplus I$ such that $\phi(\beta) = \tilde{F}\alpha_1 + \tilde{P}\alpha_2$, and choosing β_1, $\beta_2 \in D_1 \oplus D_1$ such that $\phi(\beta_i) = \alpha_i$, we have that

$$\beta = \phi^{-1}(\phi(\beta)) = (\phi^{-1} \circ \tilde{F} \circ \phi)(\beta_1) + (\phi^{-1} \circ \tilde{P} \circ \phi)(\beta_2)$$
$$\in \operatorname{Im}\bar{F} + \operatorname{Im}\bar{P}$$

Moreover, if $\alpha \in \operatorname{Im}\bar{F} \cap \operatorname{Im}\bar{P}$, then $\alpha =$

$$(\phi^{-1} \circ \tilde{F} \circ \phi)(\alpha_1) = (\phi^{-1} \circ \tilde{P} \circ \phi)(\alpha_2)$$

for some α_1, $\alpha_2 \in D_1 \oplus D_1$. Now,

$$(\tilde{F} \circ \Phi)(\alpha_1) = (0,a_1) \text{ and } (\tilde{P} \circ \Phi)(\alpha_2) = (a_2,0)$$

for some a_1, $a_2 \in I$. Since Φ^{-1} is injective, we have that $(0,a_1) = (a_2,0)$, and so, $a_1 = a_2 = 0$. It follows that

$$\alpha = (\Phi^{-1} \circ \tilde{F} \circ \Phi)(\alpha_1) = \Phi^{-1}(0,0) = (0,0)$$

This proves the claim.

Since $\bar{F}$ and $\bar{P}$ are D_1-endomorphisms of $D_1 \oplus D_1$, each is given by a 2×2 matrix (also called $\bar{F}$ and $\bar{P}$, resp.) with entries in $D_1 = (D[Y])/(g)$. Lift $\bar{F}$ and $\bar{P}$ to 2×2 matrices F and P over D[Y]—any lifting will suffice. Setting $G = [P,gI]$, where I is the 2×2 identity matrix, we claim that (F,G) is a reachable system over D[Y]. By Theorem 2.3, we can check this residually at each maximal ideal M of D[Y]. So, consider the columns of the matrix $[G,FG] = [P,gI,FP,gF]$.

If $g \notin M$, then g is a unit in (D[Y])/M and the columns of gI span $((d[Y])/M) \oplus ((D[Y])/M)$. If $g \in M$, then (D[Y])/M is a homomorphic image of $(D[Y])/(g) = D_1$. But $[\bar{G},\bar{F}\bar{G}] = [\bar{P},\bar{0}, \bar{F}\bar{P},\bar{0}]$, and since the columns of $\tilde{P}$ and $\tilde{F}\tilde{P}$ generate $I \oplus I$, the columns of $\bar{P}$ and $\bar{F}\bar{P}$ generate $D_1 \oplus D_1$. It follows readily that the columns of $[P,0,FP,0]$ generate $((D[Y])/M) \oplus ((D[Y])/M)$, where "–" denotes forming residues modulo M.

We now complete the proof of the theorem as follows, using Theorem 3.9 and its preliminaries. Since D is a one-dimensional Bézout domain, finitely generated projective D[Y]-modules are free by [6, Corollary 1]. By Theorem 3.9, if D[Y] has the pole assignability property, then ImG must contain a unimodular vector u since (F,G) is a reachable system. Because finitely generated projective D[Y]-modules are free, u is part of a D[Y]-basis {u,v} for $D[Y] \oplus D[Y]$. Going modulo g—in other words, tensoring by $(D[Y])/(g) = D_1$, we have that $\{\bar{u},\bar{v}\}$ is a basis for $D_1 \oplus D_1$ and that $\bar{u} \in \text{Im}\bar{G}$, where $\bar{G} = [\bar{P},\bar{0}]$. Thus,

$\bar{u}$ is part of a basis for $D_1 \oplus D_1$ and $\bar{u} \in \text{Im}\bar{P} \cong I \not\cong D_1$. Since I is a rank one D_1-module, we will be finished as soon as we prove the following result.

Let R be an integral domain with quotient field L and let N be a rank one submodule of $R \oplus R$ — that is, $\dim_L(N \otimes_R L) = 1$. If N contains an element which is part of a basis for $R \oplus R$, then $N \cong R$.

To justify this claim, let $\{u,v\}$ be an R-basis for $R \oplus R$ with $u \in N$. Set

$$J = N \underset{R}{:} v = \{r \in R \mid rv \in N\}$$

Then J is an ideal of R and we assert that $N = Ru \oplus Jv$. We clearly have that $N \supseteq Ru + Jv$. Since $\{u,v\}$ is an R-basis of $R \oplus R$, $Ru \cap Jv = (0)$ and if $n \in N$, $n = du + rv$ for some $d,r \in R$. But $n - du = rv \in N$ and so $r \in J$. It follows that, if $J \neq (0)$, $N \otimes_R L$ is a two-dimensional vector space, a contradiction. This proves that $J = (0)$, $N = Ru$, and that $N \cong R$. It also completes the proof of Theorem 3.20.

Theorem 3.20 has two corollaries which reveal its strength. Indeed, Theorem 3.20 arose from trying to understand why the rings $\mathbf{R}[X,Y]$, $\mathbf{C}[X,Y]$ and $Z[Y]$ do not have the pole assignability property.

THEOREM 3.21. Let k be a field with X and Y indeterminates. Then $k[X,Y]$ does not have the pole assignability property. In fact, if R is any commutative ring, then $R[X,Y]$ does not have the pole assignability property.

Proof. Let $g = Y^2 - X^3 + X \in k[X,Y]$ and set $k[x,y] = k[X,Y]/(g)$, where x and y denote the residues mod g of X and Y, respectively. We claim that $(x,y)^2$ is principal, but (x,y) is not and, moreover, that g is a prime polynomial. The last claim is easy, for $Y^2 - X^3 + X$ is an irreducible polynomial in $(k[X])[Y]$ since $X^3 - X$ isn't a square in $k[X]$. In $k[X,Y]$ irreducible polynomials are prime.

Now, $(x,y)^2 = (x^2,xy,y^2)$. Moreover, $x = -y^2 + x^3$ and so $x \in (x,y)^2$. But $y^2 = x^3 - x \in (x)$ and so $(x,y)^2 = (x)$.

So, to complete the proof we have only to prove that (x,y) is not a principal ideal of $k[x,y]$. We begin by noting that $k[x,y]$ is a simple ring extension of $k[X]$ obtained from $k[X,Y]$ by modding out by a monic polynomial of degree two. Thus, $k[x,y]$ is a free $k[X]$-module generated by 1 and y. (See Theorem 1.10.) Moreover, under the canonical homomorphism, $k[X] \cong k[x]$ and we shall regard $k[x,y]$ as a free $k[x]$-module with x an indeterminate over k.

Suppose that $(x,y) = qk[x,y]$ with $x = aq$ and $y = bq$ for $a,b \in k[x,y]$. Write $q = q_0 + q_1y$, $a = a_0 + a_1y$, and $b = b_0 + b_1y$ for $a_0, a_1, b_0, b_1, q_0, q_1 \in k[x]$. Then from $x = aq$, we have that

$$x = (a_0q_0 + a_1q_1(x^3 - x)) + (a_1q_0 + a_0q_1)y$$

from which it follows that

$$1 = [a_0q_0 + a_1q_1(x^3 - x)]/x \text{ and that } a_1q_0 + a_0q_1 = 0$$

Similarly, from $y = bq$, we have that

$$b_0q_0 + b_1q_1(x^3 - x) = 0 \text{ and } b_1q_0 + b_0q_1 = 1$$

From these equations it follows that

$$\begin{bmatrix} \frac{a_0}{x} & \frac{a_1}{x} \\ b_0 & b_1 \end{bmatrix} \begin{bmatrix} q_0 & q_1 \\ (x^3 - x)q_1 & q_0 \end{bmatrix} = \begin{bmatrix} 1 & 0 \\ 0 & 1 \end{bmatrix}$$

Taking determinants we see that

$$\left(\frac{1}{x}\right)(a_0b_1 - b_0a_1)[q_0{}^2 - q_1{}^2(x^3 - x)] = 1$$

which is the same as saying that

$$(a_0b_1 - b_0a_1)[q_0^2 - q_1^2(x^3 - x)] = x$$

If $q_1 = 0$, then $q = q_0$ and so $y = q_0b = q_0b_0 + q_0b_1y$. Thus, $q_0b_1 = qb_1 = 1$ and q is a unit. But $(q) = (x,y)$. This contradiction implies that $q_1 \neq 0$. By degree considerations it follows that, since, $q_1 \neq 0$,

$$\deg[q_0^2 - q_1^2(x^3 - x)] \geqslant 3$$

But then

$$\deg([a_0b_1 - b_0a_1][q_0^2 - q_1^2(x^3 - x)]) \geqslant 3 > 1 = \deg x$$

This contradiction completes the proof that (x,y) is not principal and that $k[X,Y]$ does not have the pole assignability property.

If R is an arbitrary commutative ring, let M be a maximal ideal of R. Then $R[X,Y]/M[X,Y] \cong (R/M)[X,Y]$ is a polynomial ring in two variables over the field R/M. By what was just proved, such a ring never has the pole assignability property. But $R[X,Y]/M[X,Y]$ is a homomorphic image of $R[X,Y]$ and, as remarked following Lemma 3.6, if $R[X,Y]$ has the pole assignability property, so does any homomorphic image. This completes the proof of Theorem 3.21.

This result coupled with Theorem 3.13 shows that if k is a field, then $k[X_1,\ldots,X_n]$ has the pole assignability property if and only if $n = 1$.

THEOREM 3.22. Let Z be the ring of integers with Y an indeterminate. Then $Z[Y]$ does not have the pole assignability property.

Proof. If $g = Y^2 + 5$, then $Z[Y]/(g) \cong Z[\sqrt{-5}]$. It is classical that $(3, \sqrt{-5} - 2)$ is a nonprincipal ideal whose square is principal.

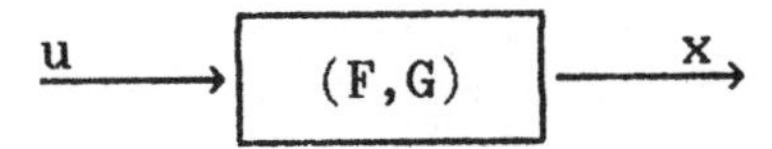

Figure 1. An open-loop system

3.4. DYNAMIC STABILIZATION

In previous sections we have seen that single input reachable systems are coefficient assignable over an arbitrary commutative ring R, but that multiple input reachable systems may fail to be coefficient assignable, or even pole assignable, unless some additional assumptions are made regarding the ring R. In this section we will show that, if the system is appropriately "enlarged," then every reachable system can be stabilized. In order to better understand the ideas here, we give the standard "black-box" formulation of linear feedback. Let (F,G) be an n-dimensional system over a commutative ring R. The system can be schematically depicted by Figure 1. In that figure, the input is u and the output is x. In the case of linear feedback, the output is fed back as the input after being transformed by the feedback matrix K. Figure 2 illustrates this.

When R is the field of real or complex numbers, the systems of differential equations corresponding to Figures 1 and 2 are

$$x' = Fx + Gu \tag{1}$$

and

$$\begin{aligned} x' &= Fx + Gu \\ u &= Kx \end{aligned} \tag{2}$$

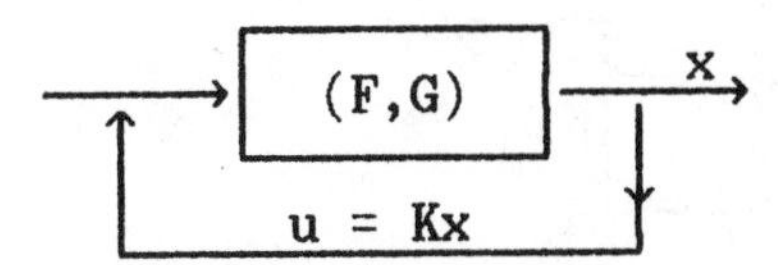

Figure 2. A linear feedback system

System (2) can be rewritten as a single equation

$$x' = (F + GK)x \tag{2'}$$

The problems of this chapter are, as we have seen earlier, concerned with the existence of an appropriate matrix K so that F + GK has some desired property.

An engineering alternative to linear (or static) feedback is to build another system (K_4, K_3) to couple to the original system and to feed back the outputs of both systems into the original system. This process is known as *dynamic feedback* and is depicted in Figure 3. When R is the field of real or complex numbers, the system of differential equations corresponding to Figure 3 is

$$\begin{aligned} x_1' &= Fx_1 + Gu \\ x_2' &= K_3x_1 + K_4x_2 \\ u &= K_1x_1 + K_2x_2 \end{aligned} \tag{3}$$

If we substitute $u = K_1x_1 + K_2x_2$ into the first equation and let

$$x = \begin{bmatrix} x_1 \\ x_2 \end{bmatrix}$$

system (3) becomes

$$x' = \left(\begin{bmatrix} F & 0 \\ 0 & 0 \end{bmatrix} + \begin{bmatrix} G & 0 \\ 0 & I \end{bmatrix} \begin{bmatrix} K_1 & K_2 \\ K_3 & K_4 \end{bmatrix} \right) x \tag{3'}$$

where the various blocks in the matrices are of appropriate

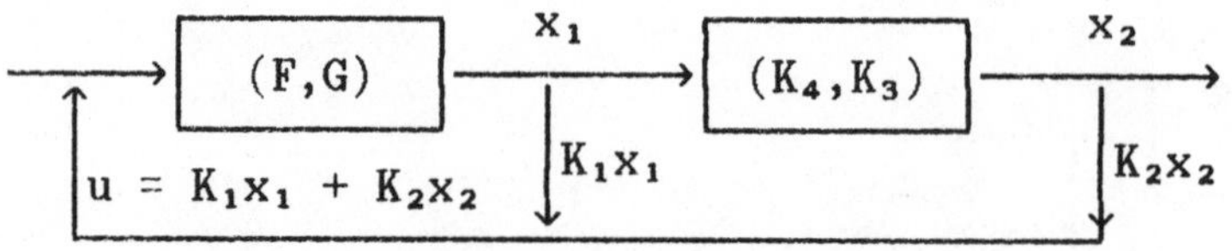

Figure 3. A dynamic feedback system

sizes. An important design problem becomes that of choosing an appropriate system (K_4,K_3) and feedback matrices K_1 and K_2 so that the matrix

$$\tilde{F} + \tilde{G}\tilde{K} = \begin{bmatrix} F & 0 \\ 0 & 0 \end{bmatrix} + \begin{bmatrix} G & 0 \\ 0 & I \end{bmatrix}\begin{bmatrix} K_1 & K_2 \\ K_3 & K_4 \end{bmatrix} = \begin{bmatrix} F + GK_1 & GK_2 \\ K_3 & K_4 \end{bmatrix}$$

has some desired property; e.g., characteristic polynomial with preassigned coefficients, characteristic polynomial with given zeros and multiplicities, or characteristic polynomial with zeros in the left half-plane of the complex number plane.

Thusly motivated, we can now define dynamic pole assignability and dynamic coefficient assignability for an arbitrary commutative ring R, and dynamic stabilizabilty for any ring of real or complex-valued functions defined on some set S. Let (F,G) be an n-dimensional system over a commutative ring R. The system (F,G) is *dynamically coefficient assignable* (resp. *dynamically pole assignable,* or *dynamically stabilizable* when R is a function ring) if there is a nonnegative integer r such that the $(n + r)$-dimensional system

$$(\tilde{F},\tilde{G}) = \left(\begin{bmatrix} F & 0 \\ 0 & 0 \end{bmatrix}, \begin{bmatrix} G & 0 \\ 0 & I \end{bmatrix}\right)$$

where I is the $r \times r$ identity matrix, is coefficient assignable (resp. pole assignable, or stabilizable when R is a function ring). Note that (F,G) is reachable if and only if $(\tilde{F},\tilde{G})$ is reachable, by Theorem 2.3 or the lemma used in the proof of Theorem 3.10.

The principal result of this section is the following nice, and at first glance, surprising result: The system (F,G) is reachable if and only if it is dynamically coefficient assignable. This is especially interesting since there are no restrictions placed on the ring R and since, in particular, it applies to all the function rings discussed in Section 3.5. From a practical viewpoint, the major difficulty occurs in the proof, where r is taken to be n^2. Thus, a (possibly) impractically large system (K_4, K_3) would have to be constructed This same number $r = n^2$ works for any ring R, but may be a very crude approximation of the smallest value of r that will work for a given ring. For example, if R is a field the minimum value of r is zero. In fact, a ring R has the coefficient assignability property if and only if, for every reachable system (F,G) over R, the minimum value of r such that the system

$$\left(\begin{bmatrix} F & 0 \\ 0 & 0 \end{bmatrix}, \begin{bmatrix} G & 0 \\ 0 & I_r \end{bmatrix}\right)$$

is coefficient assignable is zero. Determining the minimum value of r for various rings R seems to be an open problem in most cases.

In order to prove the main result of this section, we need the following technical lemma.

LEMMA 3.23. Let (F,G) be an (ns)-dimensional system over a ring R. Suppose there is an invertible matrix V and columns $\bar{g}_1, \ldots, \bar{g}_s$ of $\bar{G} = GV$ such that the square $(ns) \times (ns)$ matrix

$$\left[F^{n-1}\bar{g}_1, \ldots, \bar{g}_1, F^{n-1}\bar{g}_2, \ldots, \bar{g}_2, \ldots, F^{n-1}\bar{g}_s, \ldots, \bar{g}_s\right]$$

is invertible. Then (F,G) is systems equivalent to a system $(\hat{F} + \hat{G}K, \hat{G})$ where $\hat{F} + \hat{G}K$ has a cyclic vector in the image of $\hat{G}$. In particular, the system (F,G) is coefficient assignable.

Proof. First note that the system (F,G) is systems equivalent to the system $(F,\bar{G})$. Thus, without loss of generality, we may assume that V is the identity matrix and not bother with the "bars"; i.e., we shall assume

$$\left[F^{n-1}g_1,\ldots,g_1,F^{n-1}g_2,\ldots,g_2,\ldots,F^{n-1}g_s,\ldots,g_s\right]$$

is invertible, where each g_i is a column of G.

Next, reorder the usual basis for R^m so that g_1 is the first column of G, g_2 is the second column of G, etc. Since $[F^{n-1}g_1,\ldots,g_s]$ is invertible, $\{F^{n-1}g_1,\ldots,g_s\}$ is a basis for R^{ns}. The matrices $\tilde{F}$ and $\tilde{G}$ representing F and G with respect to these bases are

$$\tilde{F} = \begin{array}{c} \\ n\left\{\begin{matrix}\\ \\ \\ \\ \end{matrix}\right. \\ n\left\{\begin{matrix}\\ \\ \\ \\ \end{matrix}\right. \\ \vdots \\ n\left\{\begin{matrix}\\ \\ \\ \\ \end{matrix}\right. \end{array} \left[\begin{array}{cccc|cccc|c|cccc} \multicolumn{4}{c}{n} & \multicolumn{4}{c}{n} & \cdots & \multicolumn{4}{c}{n} \\ * & 1 & 0\cdots & 0 & * & 0 & \cdots & 0 & \ddots & * & & \cdots & 0 \\ * & 0 & 1 & 0 & * & 0 & & 0 & & * & & & 0 \\ \vdots & & & 1 & \vdots & & & & & \vdots & & & \\ * & 0 & 0 & 0 & * & 0 & & 0 & & * & & & 0 \\ \hline * & 0 & 0 & 0 & * & 1 & & 0 & & * & & & 0 \\ * & 0 & 0 & 0 & * & 0 & & 0 & & * & & & 0 \\ \vdots & & & & \vdots & & & 1 & & \vdots & & & \\ * & 0 & 0 & 0 & * & 0 & & 0 & & * & & & 0 \\ \hline \ddots & & & & & & & & & & & & \\ \hline * & 0 & & 0 & * & & & 0 & & * & 1 & & 0 \\ * & 0 & & 0 & * & & & 0 & & * & 0 & & \\ \vdots & & & & \vdots & & & \vdots & & \vdots & & & 1 \\ * & 0 & & 0 & * & & & 0 & & * & & & 0 \end{array}\right]$$

and

$$\tilde{G} = \begin{array}{c} \\ \\ \\ n \\ \\ \\ \\ \\ \\ n \\ \\ \\ \\ \\ \\ \\ \\ \\ n \\ \\ \\ \\ \end{array} \left[\begin{array}{cccc|ccc}
0 & 0 & \cdots & 0 & * & \cdots & * \\
0 & 0 & & 0 & * & & * \\
\vdots & & & & & & \\
0 & 0 & & 0 & * & & * \\
1 & 0 & & 0 & * & & * \\ \hline
0 & 0 & & 0 & * & & * \\
0 & 0 & & 0 & * & & * \\
\vdots & & & & & & \\
0 & 0 & & 0 & * & & * \\
0 & 1 & & 0 & * & & * \\ \hline
\cdot & & & & & & \\
& & \cdot & & & & \\
& & & \cdot & & & \\ \hline
0 & 0 & & 0 & * & & * \\
0 & 0 & & 0 & * & & * \\
\vdots & & & & & & \\
0 & 0 & & 0 & * & & * \\
0 & 0 & & 1 & * & & *
\end{array}\right]$$

$$s$$

where an * denotes a (possibly) nonzero ring element. The system $(\tilde{F},\tilde{G})$ is in a "reachability normal form." However, the form of $\tilde{F}$ is not quite simple enough for our purposes, so we will make another change of basis in R^{ns} to get a "reachability canonical form" for F. We will give the transition matrix Φ for this change of basis. In order to describe Φ, let

$$F^n g_j = - \sum_{i=1}^{s} \sum_{k=1}^{n} \alpha_{ikj} F^{n-k} g_i, \quad j = 1,\ldots,s$$

Then the $-\alpha_{ikj}$ are the elements marked by *'s in the above normal form for F. That is, column $(j - 1)n + 1$ of $\tilde{F}$ is

$$[-\alpha_{11j}, -a_{12j}, \ldots, -\alpha_{1nj}, -\alpha_{21j}, \ldots, -\alpha_{2nj}, \ldots, -\alpha_{snj}]^T$$

With this notation, we let the transition matrix Φ be the $(ns) \times (ns)$ matrix given by

$$
\Phi = \left[\begin{array}{ccccc|ccccc|c|ccccc}
1 & 0 & \cdots & & 0 & 0 & 0 & \cdots & & 0 & \cdot & 0 & \cdots & & & 0 \\
\alpha_{111} & 1 & & & 0 & \alpha_{112} & 0 & & & & & \alpha_{11s} & 0 & & & \\
\alpha_{121} & \alpha_{111} & \ddots & & 0 & \alpha_{122} & \ddots & \ddots & & & \cdot & \vdots & \ddots & \ddots & & \\
\vdots & \ddots & \ddots & \ddots & & \vdots & & \ddots & \ddots & & & \vdots & & \ddots & \ddots & \\
\alpha_{1(n-1)1} & \cdots & & \alpha_{111} & 1 & \alpha_{1(n-1)2} & \cdots & & \alpha_{112} & 0 & \cdot & \alpha_{1(n-1)s} & \cdots & & \alpha_{11s} & 0 \\
\hline
0 & 0 & \cdots & & 0 & 1 & & & & 0 & & 0 & & & & 0 \\
\alpha_{211} & 0 & & & & \alpha_{212} & \ddots & & & & & \alpha_{21s} & \ddots & & & 0 \\
\vdots & \ddots & \ddots & & & \vdots & \ddots & \ddots & & & & \vdots & & \ddots & \ddots & \\
\alpha_{2(n-1)1} & \cdots & & \alpha_{211} & 0 & \alpha_{2(n-1)2} & \cdots & & \alpha_{212} & 1 & & \alpha_{2(n-1)s} & & \cdots & & 0 \\
\hline
\cdot & & & & & & & & & & & & & & & \\
& & \cdot & & & & & & & & & & & & & \\
& & & & \cdot & & & & & & & & & & & \\
\hline
0 & 0 & \cdots & 0 & 0 & 0 & & & & 0 & & 1 & & & & 0 \\
\alpha_{s11} & 0 & & & & \alpha_{s12} & \ddots & & & & & \alpha_{s1s} & \ddots & & & 0 \\
\vdots & \ddots & \ddots & & & \vdots & & \ddots & \ddots & & & \vdots & & \ddots & \ddots & \\
\alpha_{s(n-1)1} & \cdots & & \alpha_{s11} & 0 & \alpha_{s(n-1)2} & & \cdot:\cdot & & 0 & & \alpha_{s(n-1)s} & & \cdot:\cdot & & 1
\end{array}\right]
$$

where each diagonal block has 1's on the diagonal and Toeplitz subdiagonals formed from the corresponding entries of $\tilde{F}$, and the off-diagonal blocks have 0's on the diagonal and subdiagonals as before. We claim that Φ is invertible with $\det \Phi = 1$ and that $\hat{F} = \Phi^{-1}\tilde{F}\Phi$, where $\hat{F}$ is defined to be the matrix having s^2 $n \times n$ blocks with the ijth block $\hat{F}_{ij}$ equal to

$$
\begin{bmatrix}
0 & \delta_{ij} & 0 & \cdots & 0 \\
0 & 0 & \delta_{ij} & \cdots & 0 \\
\vdots & & & & \delta_{ij} \\
-\alpha_{inj} & -\alpha_{i(n-1)j} & & \cdots & -\alpha_{ij}
\end{bmatrix}
$$

and δ_{ij} is the Kronecker delta. Notice that the form of $\hat{F}$ is similar to the form of $\tilde{F}$ except the *'s now appear in rows ni, $i = 1,\ldots,s$, instead of in columns $nj + 1$, $j = 0,\cdots,s - 1$. The fact that $\det \Phi = 1$ can be seen by induction on n and successive expansions on rows $1, n + 1,\cdots,n(s - 1) + 1$, which reduce the block sizes to $n - 1$ while preserving the form of the matrix Φ. Thus Φ is invertible. To verify that

$\hat{F} = \Phi^{-1}\tilde{F}\Phi$, we next check that $\tilde{F}\Phi = \Phi\hat{F}$ using block multiplication. Letting A_{ij} denote the ijth $n \times n$ block of an $(ns) \times (ns)$ matrix A, we wish to show that

$$\sum_{k=j}^{s} \tilde{F}_{ik}\Phi_{kj} = \sum_{k=1}^{s} \Phi_{ik}\hat{F}_{kj}$$

Now

$$\Phi_{ik}\hat{F}_{kj} = \begin{bmatrix} \delta_{ik} & 0 & 0 & \cdots & 0 \\ \alpha_{i1k} & \delta_{ik} & 0 & \cdots & 0 \\ \alpha_{i2k} & \alpha_{i1k} & \delta_{ik} & \cdots & 0 \\ \vdots & & & & \\ \alpha_{i(n-1)k} & \cdots & & \alpha_{i1k} & \delta_{ik} \end{bmatrix} \begin{bmatrix} 0 & \delta_{kj} & 0 & \cdots & 0 \\ 0 & 0 & \delta_{kj} & \cdots & 0 \\ \vdots & & & & \\ 0 & & \cdots & & \delta_{kj} \\ -\alpha_{knj} & & \cdots & & -\alpha_{k1j} \end{bmatrix}$$

$$= \begin{bmatrix} 0 & \delta_{ik}\delta_{kj} & 0 & \cdots & 0 \\ 0 & \alpha_{i1k}\delta_{kj} & \delta_{ik}\delta_{kj} & & 0 \\ 0 & \alpha_{i2j}\delta_{kj} & \alpha_{i1k}\delta_{kj} & & 0 \\ \vdots & \vdots & & & \vdots \\ \delta_{ik}(-\alpha_{knj}) & \begin{pmatrix} \alpha_{i(n-1)k}\delta_{kj} \\ + \delta_{ik}(-\alpha_{k(n-1)j}) \end{pmatrix} & \cdots & & \begin{pmatrix} \alpha_{i1k}\delta_{kj} \\ + \delta_{ik}(-\alpha_{k1j}) \end{pmatrix} \end{bmatrix}$$

and so,

$$\sum_{k=1}^{s} \Phi_{ik}\hat{F}_{kj} = \begin{bmatrix} 0 & \delta_{ij} & 0 & \cdots & 0 \\ 0 & \alpha_{i1j} & \delta_{ij} & \cdots & 0 \\ 0 & \alpha_{i2j} & \alpha_{i1j} & \cdots & 0 \\ \vdots & & & & \\ -\alpha_{inj} & 0 & 0 & \cdots & 0 \end{bmatrix}$$

Similarly,

$$\tilde{F}_{ik}\Phi_{kj} = \begin{bmatrix} -\alpha_{i1k} & \delta_{ik} & 0 & \cdots & 0 \\ -\alpha_{i2k} & 0 & \delta_{ik} & \cdots & 0 \\ -\alpha_{i3k} & 0 & 0 & \cdots & 0 \\ \vdots & & & & \\ -\alpha_{i(n-1)k} & 0 & 0 & \cdots & \delta_{ik} \\ -\alpha_{ink} & 0 & 0 & \cdots & 0 \end{bmatrix} \begin{bmatrix} \delta_{kj} & 0 & 0 & \cdots & 0 \\ \alpha_{k1j} & \delta_{kj} & 0 & \cdots & 0 \\ \alpha_{k2j} & \alpha_{k1j} & \delta_{kj} & \cdots & 0 \\ \vdots & & & & \\ \alpha_{k(n-1)j} & \cdots & & \alpha_{k1j} & \delta_{kj} \end{bmatrix}$$

$$= \begin{bmatrix} \begin{pmatrix} -\alpha_{i1k}\delta_{kj} \\ +\delta_{ik}\alpha_{k1j} \end{pmatrix} & \delta_{ik}\delta_{kj} & 0 & \cdots & 0 \\ \begin{pmatrix} -\alpha_{i2k}\delta_{kj} \\ +\delta_{ik}\alpha_{k2j} \end{pmatrix} & \delta_{ik}\alpha_{k1j} & \delta_{ik}\delta_{kj} & \cdots & 0 \\ \vdots & & & & \\ \begin{pmatrix} -\alpha_{i(n-1)k}\delta_{kj} \\ +\delta_{ik}\alpha_{k(n-1)j} \end{pmatrix} & \delta_{ik}\alpha_{k(n-2)j} & \cdots & \delta_{ik}\alpha_{k1j} & \delta_{ik}\delta_{kj} \\ -\alpha_{ink}\delta_{kj} & 0 & \cdots & 0 & 0 \end{bmatrix}$$

and hence,

$$\sum_{k=1}^{s} \tilde{F}_{ik}\Phi_{kj} = \begin{bmatrix} 0 & \delta_{ij} & 0 & \cdots & 0 \\ 0 & \alpha_{i1j} & \delta_{ij} & \cdots & 0 \\ \vdots & & & & \\ 0 & \alpha_{i(n-2)j} & \cdots & \alpha_{i1j} & \delta_{ij} \\ -\alpha_{inj} & 0 & \cdots & 0 & 0 \end{bmatrix}$$

$$= \sum_{k=1}^{s} \Phi_{ik}\hat{F}_{kj}$$

Since we have just made a type (1) systems transformation, $\tilde{G}$ goes to $\hat{G} = \Phi^{-1}\tilde{G}$. However, since $\Phi\hat{G} = \tilde{G}$ and $\Phi e_{nj} = e_{nj}$, where

$$e_{nj} = [0,\ldots,\underset{nj}{1},0,\ldots,0]^T$$

is the usual basis element in R^{ns}, we see that the first s columns of $\hat{G}$ are the same as the first s columns of $\tilde{G}$. Hence, $\hat{G}$ has exactly the same form as does $\tilde{G}$.

We now claim that we can find a matrix K and a vector u such that $\hat{G}u$ is cyclic for $\hat{F} + \hat{G}K$. Indeed, consider $\hat{G}K$, where K is an $m \times (ns)$ matrix to be determined. By the form of $\hat{G}$, rows $n, 2n, \ldots, sn$ of $\hat{G}K$ are exactly the same as rows $1, 2, \ldots, s$ of K if we take all rows of K to be zero after the sth row. Then, by suitably choosing the first s rows of K, we can make $\hat{F} + \hat{G}K$ into the matrix whose first superdiagonal consists entirely of ones and that has zeros everywhere else. The vector $\hat{G}e_s = e_{ns}$ is clearly cyclic for $\hat{F} + \hat{G}K$.

The final assertion follows from Theorem 3.3 and the fact that coefficient assignability is preserved under systems equivalence.

THEOREM 3.24. Let (F,G) be an n-dimensional system over a ring R. Then the following are equivalent.

1. (F,G) is dynamically pole assignable.
2. (F,G) is reachable.
3. There exists a positive integer r such that the system

$$\left(\begin{bmatrix} F & 0 \\ 0 & 0 \end{bmatrix}, \begin{bmatrix} G & 0 \\ 0 & I_r \end{bmatrix}\right) = (\tilde{F}, \tilde{G})$$

"feeds back to a cyclic vector"—that is, there exists a matrix $\tilde{K}$ and a vector u such that $\tilde{G}u$ is a cyclic vector for $\tilde{F} + \tilde{G}\tilde{K}$.

4. (F,G) is dynamically coefficient assignable.

Proof. (1) ⟹ (2): Assume that (F,G) is dynamically pole assignable. Then there is a number r such that the (n + r)-dimensional system $(\tilde{F},\tilde{G})$, where

$$\tilde{F} = \begin{bmatrix} F & 0 \\ 0 & 0 \end{bmatrix} \quad \text{and} \quad \tilde{G} = \begin{bmatrix} G & 0 \\ 0 & I \end{bmatrix}$$

and I is the r × r identity matrix, is pole assignable. By Theorem 3.1, $(\tilde{F},\tilde{G})$ is reachable, and by Theorem 2.3, (F,G) is reachable, since the columns of $[G,FG,\ldots,F^{n-1}G]$ generate R^n if and only if the columns of $[\tilde{G},\tilde{F}\tilde{G},\ldots,\tilde{F}^{n+r-1}\tilde{G}]$ generate R^{n+r}.

(2) ⟹ (3): Assume that (F,G) is reachable so that the matrix $[G,FG,\ldots,F^{n-1}G]$ has a right inverse C over R. Let

$$\tilde{F} = \begin{bmatrix} F & 0 \\ 0 & 0 \end{bmatrix} \quad \text{and} \quad \tilde{G} = \begin{bmatrix} G & 0 \\ 0 & I \end{bmatrix}$$

where I is the $n^2 \times n^2$ identity matrix.

We claim that there are matrices K and L such that the $(n + n^2)$-dimensional system $(\bar{F},\bar{G}) = (\tilde{F} + \tilde{G}K,\tilde{G}L)$ satisfies the hypotheses of Lemma 3.23. To prove the claim, partition the nm × n matrix C so that C_1 is the first m rows, C_2 is the second m rows, etc; i.e.,

$$C = \begin{bmatrix} C_1 \\ \vdots \\ C_n \end{bmatrix}$$

where each C_i is an m × n matrix. Let K be an $(m + n^2) \times (n + n^2)$ matrix defined as follows:

$$K = \begin{bmatrix} K_1 & K_2 \\ K_3 & K_4 \end{bmatrix}$$

where K_1 is the m × n zero matrix, K_3 is the $n^2 \times n$ zero matrix, K_4 is the $n^2 \times n^2$ matrix partitioned into n × n blocks with $n \times n$ identity matrices on the first (block) superdiagonal and zeros elsewhere, and K_2 is the $m \times n^2$ block matrix $[C_1\ C_2\ \cdots\ C_n]$. Then

$$K = \left[\begin{array}{c|cccccc} 0 & C_1 & C_2 & C_3 & \cdots & C_n \\ \hline 0 & 0 & I & 0 & \cdots & 0 \\ 0 & 0 & 0 & I & \cdots & 0 \\ \vdots & \vdots & & & & \\ 0 & 0 & 0 & 0 & \cdots & I \\ 0 & 0 & 0 & 0 & \cdots & 0 \end{array}\right]$$

Next, let

$$L = \begin{bmatrix} 0 \\ 0 \\ \vdots \\ 0 \\ I \end{bmatrix}$$

where L is an $(m + n^2) \times n$ matrix. In block form, the first block is the $m \times n$ zero matrix and all other blocks are $n \times n$. Then

$$\bar{F} = \tilde{F} + \tilde{G}K = \begin{bmatrix} F & 0 \\ 0 & 0 \end{bmatrix} + \begin{bmatrix} G & 0 \\ 0 & I \end{bmatrix} \begin{bmatrix} K_1 & K_2 \\ K_3 & K_4 \end{bmatrix}$$

$$= \begin{bmatrix} F & 0 \\ 0 & 0 \end{bmatrix} + \begin{bmatrix} GK_1 & GK_2 \\ K_3 & K_4 \end{bmatrix}$$

$$= \begin{bmatrix} F & 0 \\ 0 & 0 \end{bmatrix} + \begin{bmatrix} 0 & GK_2 \\ 0 & K_4 \end{bmatrix} = \begin{bmatrix} F & GK_2 \\ 0 & K_4 \end{bmatrix}$$

$$= \begin{bmatrix} F & GC_1 & GC_2 & \cdots & GC_n \\ 0 & 0 & I & \cdots & 0 \\ \vdots & \vdots & & & \\ 0 & 0 & 0 & \cdots & I \\ 0 & 0 & 0 & \cdots & 0 \end{bmatrix}$$

and

$$\bar{G} = \tilde{G}L = \begin{bmatrix} G & 0 & 0 & \cdots & 0 \\ 0 & I & 0 & \cdots & 0 \\ \vdots & \vdots & & & \\ 0 & 0 & 0 & \cdots & I \end{bmatrix} \begin{bmatrix} 0 \\ 0 \\ \vdots \\ I \end{bmatrix} = \begin{bmatrix} 0 \\ 0 \\ \vdots \\ I \end{bmatrix}$$

where all blocks in $\bar{G}$ are $n \times n$. We next calculate $[\bar{G}, \bar{F}\bar{G}, \ldots, \bar{F}^n\bar{G}]$ by successively calculating $\bar{F}\bar{G}, \ldots, \bar{F}^n\bar{G}$.

$$\bar{F}\bar{G} = \begin{bmatrix} F & GC_1 & GC_2 & \cdots & GC_n \\ 0 & 0 & I & \cdots & 0 \\ \vdots & & & & \\ 0 & 0 & 0 & \cdots & I \\ 0 & 0 & 0 & \cdots & 0 \end{bmatrix} \begin{bmatrix} 0 \\ 0 \\ \vdots \\ \\ I \end{bmatrix} = \begin{bmatrix} GC_n \\ 0 \\ \vdots \\ I \\ 0 \end{bmatrix}$$

$$\bar{F}^2\bar{G} = \bar{F}(\bar{F}\bar{G}) = \begin{bmatrix} F & GC_1 & GC_2 & \cdots & GC_n \\ 0 & 0 & I & \cdots & 0 \\ \vdots & & & & \\ 0 & 0 & 0 & \cdots & I \\ 0 & 0 & 0 & \cdots & 0 \end{bmatrix} \begin{bmatrix} GC_n \\ 0 \\ \vdots \\ I \\ 0 \end{bmatrix} = \begin{bmatrix} FGC_n + GC_{n-1} \\ \vdots \\ I \\ 0 \\ 0 \end{bmatrix}$$

etc.

We finally get

$$[\bar{G}, \bar{F}\bar{G}, \ldots, \bar{F}^n\bar{G}] = \begin{bmatrix} 0 & GC_n & FGC_n + GC_{n-1} & \cdots & F^{n-1}GC_n + \cdots + GC_1 \\ 0 & 0 & 0 & \cdots & 0 \\ \vdots & & & & \\ 0 & I & 0 & \cdots & 0 \\ I & 0 & 0 & \cdots & 0 \end{bmatrix}$$

$$= \begin{bmatrix} 0 & * & * & \cdots & * & I \\ 0 & 0 & 0 & \cdots & I & 0 \\ \vdots & & & & & \\ 0 & I & 0 & \cdots & 0 & 0 \\ I & 0 & 0 & \cdots & 0 & 0 \end{bmatrix}$$

since

$$GC_1 + FGC_2 + \cdots + F^{n-1}GC_n = [G, GF, \ldots, F^{n-1}G] \begin{bmatrix} C_1 \\ C_2 \\ \vdots \\ C_n \end{bmatrix} = I$$

(The *'s denote known, possibly nonzero, $n \times n$ matrices.) Hence, we see that $[\bar{G}, \bar{F}\bar{G}, \ldots, \bar{F}^n\bar{G}]$ is an $(n + n^2) \times (n + n^2)$ invertible matrix, and thus, $(\bar{F}, \bar{G})$ satisfies the hypotheses of the lemma with $s = n$ and $V = I$. This establishes the claim. The conclusion of the lemma then shows that (2) ⟹ (3).

(3) ⟹ (4): This follows from Theorem 3.3.

(4) ⟹ (1): This is trivial.

Note that, in the proof of (2) ⟹ (3), $(\bar{F}, \bar{G}) = (\tilde{F} + \tilde{G}K, \tilde{G}L)$, where

$$K = \begin{bmatrix} K_1 & K_2 \\ K_3 & K_4 \end{bmatrix} = \begin{bmatrix} 0 & K_2 \\ 0 & K_4 \end{bmatrix} \text{ and } L = \begin{bmatrix} L_1 \\ L_2 \end{bmatrix} = \begin{bmatrix} 0 \\ L_2 \end{bmatrix}$$

(Both K_1 and L_1 are the $m \times n$ zero matrix.) By Theorem 3.3, for a given monic polynomial p of degree $n + n^2$, there is a matrix $\bar{K}$ such that $\bar{F} + \bar{G}\bar{K}$ has characteristic polynomial p. But

$$\bar{F} + \bar{G}\bar{K} = \tilde{F} + \tilde{G}K + \tilde{G}L\bar{K} = \tilde{F} + \tilde{G}(K + L\bar{K})$$

and hence, $(\tilde{F}, \tilde{G})$ is coefficient assignable with feedback matrix $K + L\bar{K}$. Writing the $n \times (n + n^2)$ matrix $\bar{K}$ in the form

$$\bar{K} = [\bar{K}_1 \quad \bar{K}_2]$$

where $\bar{K}_1$ is $n \times n$ and $\bar{K}_2$ is $n \times n^2$, we see that

$$K + L\bar{K} = \begin{bmatrix} 0 & K_2 \\ 0 & K_4 \end{bmatrix} + \begin{bmatrix} 0 & 0 \\ L_2\bar{K}_1 & L_2\bar{K}_2 \end{bmatrix} = \begin{bmatrix} 0 & K_2 \\ L_2\bar{K}_1 & K_4 + L_2\bar{K}_2 \end{bmatrix}$$

Thus, $K + L\bar{K} = \tilde{K}$ is of the form

$$\begin{bmatrix} 0 & \tilde{K}_2 \\ \tilde{K}_3 & \tilde{K}_4 \end{bmatrix}$$

This says that there was no direct feedback of the output x_1 of the first "black-box" in the coupled system shown in Figure 3. This is illustrated in Figure 4.

When R is the field of real or complex numbers, the system of differential equations corresponding to Figure 4, after substituting in $u = \tilde{K}_2 x_2$ and using

$$x = \begin{bmatrix} x_1 \\ x_2 \end{bmatrix}$$

is

$$x' = \begin{bmatrix} F & G\tilde{K}_2 \\ \tilde{K}_3 & \tilde{K}_4 \end{bmatrix} x \tag{4}$$

More restrictive definitions of dynamic properties could be given in terms of the matrix

$$\begin{bmatrix} F & G\tilde{K}_2 \\ \tilde{K}_3 & \tilde{K}_4 \end{bmatrix}$$

from equation (4). In particular, we could define *restricted dynamic coefficient assignability* as follows: The n-dimensional system (F,G) is restrictedly dynamically coefficient assignable if and only if there is a nonnegative integer s such that, for any given monic polynomial p of degree n + s, the characteristic polynomial of the $(n + s) \times (n + s)$ matrix

$$\begin{bmatrix} F & G\tilde{K}_2 \\ \tilde{K}_3 & \tilde{K}_4 \end{bmatrix}$$

is p for some choice of $\tilde{K}_2$, $\tilde{K}_3$ and $\tilde{K}_4$.

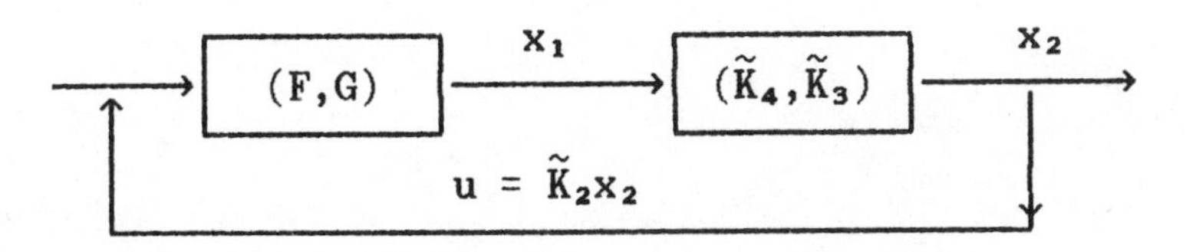

Figure 4. Restricted dynamic feedback

Clearly, if the system (F,G) is restrictedly dynamically coefficient assignable for some integer s, then (F,G) is dynamically coefficient assignable for r = s using the feedback matrix

$$K = \begin{bmatrix} 0 & \tilde{K}_2 \\ \tilde{K}_3 & \tilde{K}_4 \end{bmatrix}$$

Also, if the original system (F,G) is coefficient assignable, so that r = 0, we cannot take s = 0 since that would require det(XI − (F + GO)) to be any polynomial we want. In general, the minimum value of s for a given ring R will be larger than the corresponding minimum value for r using unrestricted dynamic feedback. The problem of determining the minimum value of s seems to be an open problem as is the relation between the minimum values using these two definitions.

3.5 PARAMETRIC STABILIZATION

For F a complex n × n matrix, consider the differential equatio[n]

$$x'(t) = Ax(t), \quad t \geq 0 \qquad (*)$$

where x(t) is an n-tuple of functions. We say that this differ[-]ential equation is *stable* if the solution x(t) satisfies

$$\lim_{t\to\infty} \|x(t)\| = 0$$

(This condition is sometimes called asymptotically stable.) Since the solutions of (*) are given by $x(t) = e^{At}x(0)$, it follows that the differential equation is stable if and only if

$$\lim_{t\to\infty} e^{At} = 0$$

But

$$\lim_{t \to \infty} e^{At} = 0$$

if and only if all the eigenvalues of A have negative real parts. See [3, Chapter 4], [4, Section 3.1] or [21, Chapter V, Section 6] for a discussion of these matters. We will say that a complex matrix A is a *stable matrix* if all the eigenvalues of A have negative real parts.

Let (H,F,G) be a classical system over the complex numbers, where H is $p \times n$, F is $n \times n$ and G is $n \times m$. That is,

$$\frac{dx}{dt} = Fx(t) + Gu(t)$$

$$y(t) = Hx(t)$$

A classical control theory problem is to find a feedback matrix K so that F − GK is a stable matrix. Letting $u(t) = -Kx(t)$, we then have a new system

$$\frac{dx}{dt} = (F - GK)x(t)$$

so that the solutions to this system go to zero at infinity.

In this section we will be concerned with systems (H,F,G) whose entries depend on a parameter. That is, let R be a ring of complex-valued functions defined on a set S and let (H,F,G) be a system over R. For $A = (a_{ij})$ a matrix over R and s in S let A(s) be the complex matrix $(a_{ij}(s))$. Assume that each system (H(s), F(s), G(s)) is reachable so that (since fields have the pole assignability property) for each s there is a complex matrix K_s such that $F(s) - G(s)K_s$ is a stable matrix. We want to find conditions on R that imply the existence of a matrix K <u>over R</u> such that, for each s in S, $F(s) - G(s)K(s)$ is a stable matrix. Before we do this we need some results from classical control theory.

Recall that, if x and y are vectors in C^n, their inner product is given by

$$(x,y) = \sum_{i=1}^{n} x_i \bar{y}_i$$

Let $M_n(C)$ denote the algebra of $n \times n$ complex matrices. A matrix $Q \in M_n(C)$ is called *positive*, written $Q \geq 0$, if $(Qx,x) \geq 0$ for all $x \in C^n$. A positive matrix Q is automatically hermitian; that is $Q = Q^*$, where Q^* denotes the conjugate transpose of Q.

THEOREM 3.25. Let $A \in M_n(C)$ be a stable matrix and define a linear map $T : M_n(C) \longrightarrow M_n(C)$ by $T(X) = AX + XA^*$. Then T is invertible, and if $Q \in M_n(C)$ is positive, then the unique P such that $T(P) = Q$ satisfies $-P \geq 0$. Moreover, P is invertible if and only if the pair (A,Q) is reachable.

Proof. Since $M_n(C)$ is finite dimensional, to prove that T is invertible we need only prove that T is one-to-one. So assume that $T(X) = AX + XA^* = 0$. Then, for this X, we have

$$\frac{d}{dt}\left(e^{At}Xe^{A^*t}\right) = e^{At}(AX + XA^*)e^{A^*t} = 0$$

so that $e^{At}Xe^{A^*t}$ is constant. Hence, $e^{At}Xe^{A^*t} = X$ for all $t \geq 0$. But

$$\lim_{t\to\infty} e^{At} = 0$$

since A is a stable matrix. Since A^* is also stable, it follows that $X = 0$. Thus, T is invertible.

If $Q \geq 0$, let

$$P = -\int_0^\infty e^{At}Qe^{A^*t}\,dt$$

Since A is a stable matrix, there is an $m > 0$ and an $\varepsilon > 0$ such that

$$\|e^{At}\| \leq me^{-\varepsilon t}$$

for all $t \geq 0$, where $\| \ \|$ denotes the operator norm of A; see [3, Chapter 4]. It follows that the improper integral defining P is norm convergent. Then

$$\begin{aligned} AP + PA^* &= -\int_0^\infty e^{At}(AQ + QA^*)e^{A^*t}\, dt \\ &= -\int_0^\infty \frac{d}{dt}\,(e^{At}Qe^{A^*t})\, dt \\ &= -(\lim_{t\to\infty} e^{At}Qe^{A^*t} - Q) = Q \end{aligned}$$

It follows from the definition of P that $-P$ is positive.

If $Px = 0$, then $0 = (Px,x)$ so

$$\int_0^\infty \|Q^{1/2}e^{A^*t}x\|^2\, dt = 0$$

Here we have used the fact that a positive matrix has a square root. (This can be seen by diagonalizing Q and taking the square root of the diagonal entries.) It follows that $Qe^{A^*t}x = 0$ for all $t \geq 0$. Looking at the coefficients of the power series of Qe^{A^*t}, we see that $QA^{*j}x = 0$ for all $j \geq 0$. Conversely, if $QA^{*j}x = 0$ for all j, then $Px = 0$. Hence $Px = 0$ if and only if $QA^{*j}x = 0$ for all j. But the definition of (Q,A^*) being observable was that $QA^{*j}x = 0$ for all $j \geq 0$ implies $x = 0$. Thus, P is invertible if and only if (Q,A^*) is observable. Since (Q,A^*) is observable if and only if (A,Q) is reachable (exercise), this completes the proof.

We also need the following form of a converse to Theorem 3.25.

LEMMA 3.26. Let $A,B,C \in M_n(C)$ with B positive and invertible and let $\alpha \geq 0$ be given. If $AB + BA^* = -\alpha CC^*$, then all the eigenvalues of A have real parts less than or equal to zero.

Proof. Suppose $A^*x = \lambda x$ for $0 \neq x \in C^n$. Then

$$\begin{aligned} 0 \geq -\alpha(CC^*x,x) &= (ABx,x) + (BA^*x,x) \\ &= (Bx,\lambda x) + (B\lambda x,x) \\ &= (\lambda + \bar{\lambda})(Bx,x) \end{aligned}$$

Since $(Bx,x) > 0$ it follows that $(\lambda + \bar{\lambda}) \leq 0$.

We can now prove the main result of this section. After we state and prove the theorem, we will give examples of function rings that satisfy the hypotheses.

THEOREM 3.27. Let R be a ring of complex-valued functions defined on a set S. Assume that R satisfies the following two conditions:

1. If $p \in R$ and $p(s) \neq 0$ for all $s \in S$, then there is a $q \in R$ with $1 \leq p(s)q(s)$ for all $s \in S$.
2. If $A \in M_n(R)$, then there is a $w \in R$ such that $w(s) \geq 1$ for all $s \in S$ and the eigenvalues of $A(s) + w(s)I$ are in the open right half-plane for all s in S.

Then, if (H,F,G) is a system over R such that $(F(s),G(s))$ is reachable for all $s \in S$, there is a matrix K over R such that $F(s) - G(s)K(s)$ is a stable matrix for each $s \in S$.

Proof. By hypothesis (2), there is a $w \in R$ such that $w(s) \geq 1$ and $-(F(s) + w(s)I)$ is a stable matrix for all $s \in S$. Define an R-linear map L from $M_n(R)$ to $M_n(R)$ by

$$L(X) = (F + wI)X + X(F + wI)^*$$

For $s \in S$, let $T_s : M_n(C) \longrightarrow M_n(C)$ be defined by

$$T_s(\theta) = (F(s) + w(s)I)\theta + \theta(F(s) + w(s)I)^*$$

By Theorem 3.25 each T_s is invertible. Let $E_{ij} \in M_n(R)$ be the matrix with 1 in the (i,j)-entry and zeros elsewhere and let

e_{ij} be the corresponding matrix in $M_n(C)$. Clearly $E_{ij}(s) = e_{ij}$ for each $s \in S$. It follows from the definitions of L and T_s that $L(X)(s) = T_s(X(s))$ for all $X \in M_n(R)$ and all $s \in S$. It is then clear that the matrix of T_s with respect to the basis $\{e_{ij}\}$ is the matrix of L with respect to the R-basis $\{E_{ij}\}$ evaluated at s. Let $d \in R$ be the determinant of L. Then d(s) is the determinant of T_s. Hence, since T_s is invertible, $d(s) \neq 0$ for all $s \in S$.

Let $N : M_n(R) \longrightarrow M_n(R)$ be the R-linear map induced by the cofactor matrix of the matrix of L. Then $N \circ L = L \circ N = dI$. Let $X = N(GG^*)$. We claim that $(1/d)X$ is an hermitian matrix. Indeed,

$$\begin{aligned} L(X) &= (L \circ N)(GG^*) \\ &= dGG^* = (F + wI)X + X(F + wI)^* \end{aligned} \tag{1}$$

For each $s \in S$, we then have that

$$d(s)GG^*(s) = (F(s) + w(s)I)X(s) + X(s)(F(s) + w(s)I)^* \tag{2}$$

Since $-(F(s) + w(s)I)$ is a stable matrix and $GG^*(s)$ is positive, it follows from Theorem 3.25 applied to $-(F(s) + w(s)I)$ and $-(1/d(s))X(s)$ that $(1/d(s))X(s)$ is positive. Hence, $(1/d)X$ is an hermitian matrix.

Now let Z be the cofactor matrix of X, let $K = G^*Z$, and let $p = 2\det(X)$. By hypothesis $(F(s),G(s))$ is reachable, so it is immediate that $(F(s) + w(s)I, G(s))$ is reachable. We claim that $(F(s) + w(s)I,(GG^*)(s))$ is also reachable. Indeed, suppose $x \in M_n(C)$ is such that $(GG^*)(s)(F(s) + w(s)I)^{*j}x = 0$ for all $j \geq 0$. Then, taking inner products, we have that for all $j \geq 0$

$$0 = ((GG^*)(s)(F(s) + w(s)I)^{*j}x,(F(s) + w(s)I)^{*j}x)$$

or

$$0 = (G^*(s)(F(s) + w(s)I)^{*j}x,G^*(s)(F(s) + w(s)I)^{*j}x)$$

Thus, $G^*(s)(F(s) + w(s)I)^{*j}x = 0$ for all $j \geq 0$ and x must be zero since by duality $(G^*(s),(F(s) + w(s)I)^*)$ is observable. Hence $(F(s) + w(s)I,(GG^*)(s))$ is reachable. Now from equation (2), we see that $(1/d(s))X(s)$ is invertible for all s by using Theorem 3.25. Hence $p(s) \neq 0$ all $s \in S$.

By condition (1), there is a $q \in R$ with $1 \leq p(s)q(s)$ for all $s \in S$. Note that this implies that $p(s)q(s)$ is real. Since Z is the cofactor matrix of X, $ZX = XZ = (p/2)I$ and $Z(s)X(s) = X(s)Z(s) = p(s)/2$. Then $Z(s)(X(s)/d(s)) = (p(s)/2d(s))I$ and $(X(s)/d(s))$ is hermitian, so that $(X(s)/d(s))Z(s)^* = (\overline{p(s)}/2\overline{d(s)})I$, or $X(s)Z(s)^* = (d(s)\bar{p}(s)/2\bar{d}(s))I$. We then compute

$$\begin{aligned}(F + wI - dqGK)X &+ X(F + wI - dqGK)^* \\ &= dGG^* - dqGG^*ZX - \bar{d}\bar{q}XZ^*GG^* \\ &= dGG^* - dq(p/2)GG^* - \bar{d}\bar{q}(d\bar{p}/2\bar{d})GG^* \\ &= dGG^* - dq(p/2)GG^* - (dpq/2)GG^* \\ &= (d - dpq)GG^*\end{aligned}$$

where the last equality is true because pq is real. Since $1 \leq p(s)q(s)$ for all s and $(1/d(s))X(s)$ is positive and invertible, it follows from Lemma 3.26 that the eigenvalues of $(F + wI - dqGK)(s)$ have real part less than or equal to zero. Hence, the eigenvalues of $F(s) - G(s)(d(s)q(s)K(s))$ all have real parts less than or equal to -1. This completes the proof of the theorem.

We now give some examples of function rings that satisfy the hypotheses of Theorem 3.27. The most interesting example (the example that is hardest to prove by other methods) is the ring R of polynomials in n variables over the real numbers. This is the ring in systems theory that arises when one considers delay differential systems in more than one delay. Consider R as a ring of functions over $S = \mathbb{R}^n$. For $A \in M_n(R)$ let

$$w = 1 + \sum_{i,j=1}^{n} a_{ij}^2$$

If $\| \ \|$ denotes the operator norm of a matrix, we then have that for $s \in \mathbf{R}^n$

$$\|A(s)\| \leq \left[\sum_{i,j=1}^{n} a_{ij}(s)^2\right]^{1/2} < w(s)$$

Since the spectral radius of A(s) is less than or equal to the operator norm (see [3, p. 6], for instance) it follows that all eigenvalues of $A(s) + w(s)I$ are in the open right half-plane, whence condition (2) is satisfied. In order to show that R satisfies condition (1), we use the real nullstellensatz from [17]. This theorem implies that if $p \in R$ is such that $p(s) \neq 0$ for all $s \in \mathbf{R}^n$, then there are positive numbers α_i and rational functions u_i such that $1 + \sum\alpha_i u_i^2 = pq$ for some $q \in R$. Hence, $1 \leq p(s)q(s)$ for all $s \in \mathbf{R}^n$ and condition (1) is satisfied.

Other examples are the rings of real or complex-valued continuous functions defined on any topological space S. In these cases, if $p(s) \neq 0$ for all $s \in S$, then $q(s) = (1/p(s))$ is also in the ring so that condiition (1) is satisfied. For any $A \in M_n(R)$ we can satisfy condition (2) by taking

$$w = 1 + \sum_{i,j=1}^{n} |a_{ij}|^2$$

Finally, the hypotheses of Theorem 3.27 are satisfied if R is the disc algebra of all complex-valued functions continuous on the closed disc and analytic on the open disc, or if R is the algebra of absolutely convergent Fourier series on the circle; see [50] for a discussion of these algebras. These algebras are inverse closed and hence satisfy condition (1). If $A \in M_n(R)$, then, since the entries are bounded functions, there is a number $\alpha > 0$ such that $\|A(s)\| < \alpha$ for all s and we can choose the w in condition (2) to be α. See [24, Sections 1 and 5] for applications of systems defined over the algebra of absolutely convergent Fourier series.

3.6 STATE ESTIMABILITY

We conclude this chapter by briefly discussing a problem dual to coefficient assignability. If (H,F,G) is a system over a ring R, we say that (H,F,G) is *state estimable* if and only if, for any monic polynomial p(X), there exists an $n \times p$ matrix L such that $\det(XI - (F - LH)) = p(X)$. In analogy with coefficient assignability, we ask

> For which rings is it true that all observable systems are state estimable?

It is classical that a field has the property, but the following example from [14] shows that the ring of integers does not.

Consider the system

$$\left([1,-1], \begin{bmatrix} 1 & 0 \\ 1 & 2 \end{bmatrix}, \begin{bmatrix} 1 \\ 2 \end{bmatrix} \right)$$

over the ring Z. Then

$$[G,FG] = \begin{bmatrix} 1 & 1 \\ -2 & -3 \end{bmatrix} \text{ and } [H^T, F^TH^T] = \begin{bmatrix} 1 & 0 \\ -1 & -2 \end{bmatrix}$$

Thus, the system is observable. It is easy to verify that there do not exist integers a and b so that the characteristic polynomial of

$$F - \begin{bmatrix} a \\ b \end{bmatrix} H$$

equals X^2. Hence the system isn't state estimable, even in the weaker sense of requiring only that the roots of the characteristic polynomial be modifiable.

The following results extend the "classical" theory to the case of noetherian rings that are equal to their own total quotient rings. Recall that such rings have only a finite number of maximal ideals. See Theorem 1.7 with M = R.

THEOREM 3.28. Let R be a noetherian ring that is equal to its own total quotient ring. A system (H,F,G) is observable over R if and only if it is state estimable over R.

Proof. By Theorem 2.9, (H,F,G) is observable if and only if the dual system (G^T,F^T,H^T) is reachable. Now since R has only finitely many maximal ideals, Theorem 3.7 applies. Therefore, (G^T,F^T,H^T) is reachable if and only if it is coefficient assignable, if and only if given a monic polynomial p(X) over R, there exists a $p \times n$ matrix K such that

$$\begin{aligned} p(X) &= \det([XI - (F^T - H^TK)]) = \det(XI - (F - K^TH)^T]) \\ &= \det([XI - (F - K^TH)]) \end{aligned}$$

if and only if the system (H,F,G) is state estimable.

There is one additional bit of terminology. Given a system (H,F,G) over a ring R, if (H,F,G) is both coefficient assignable and state estimable, then we say that the "regulator problem is solvable for (H,F,G)."

THEOREM 3.29. Let R be a noetherian ring that is equal to its own total quotient ring. The regulator problem is solvable for each system over R that is both reachable and observable.

Proof. This follows from Theorems 3.7 and 3.28.

EXERCISES

1. Let R be a ring with the pole assignability (resp. coefficient assignability) property. Let I be an ideal of R. Prove

that R/I has the pole assignability (resp. coefficient assignability) property.

2. Let $I_1, I_2, \ldots, I_n$ be ideals in a ring R. If

$$I_i + I_j = R, \text{ for all } i \neq j$$

prove that

$$I_1 + \prod_{j=2}^{n} I_j = R$$

3. Let R be a ring and let $I_1, I_2, \ldots, I_n$ be ideals such that

$$I_i + I_j = R, \text{ for all } i \neq j$$

Given elements $x_1, x_2, \ldots, x_n \in R$, prove there exists $x \in R$ such that $x - x_i \in I_i$ for all i. (Hint: Use induction on n and the previous exercise. For $n = 2$, if $1 = a_1 + a_2$, $a_i \in I_i$, take $x = x_2 a_1 + x_1 a_2$.) (This is the Chinese remainder theorem.)

4. Use the previous exercise to prove that if R has only finitely many maximal ideals, then R/J is a direct sum of fields, where J is the Jacobson radical of R.

5. Let B be a matrix of any size with entries in a ring R. Suppose that all 2×2 minors of B are zero. If F and G are matrices of the appropriate sizes, prove that all 2×2 minors of BF and GB are zero.

6. Show that if $v \in R^m$ is unimodular, G is any $m \times n$ matrix with entries in R, $u \in R^n$ and $v = Gu$, then u is unimodular.

7. Let A be an $n \times n$ (*)-matrix over a ring R. Prove that the characteristic polynomial of A is of the form

$$X^n + \sum_{j=0}^{n-1} a_j X^j$$

where $a_0, a_1, \ldots, a_{n-2}$ are in the Jacobson radical of R.

8. Prove directly that if G is a good matrix over a ring R, then the ideal of R generated by the entries of G is R.

9. Prove that, for P and G any two matrices over a ring R, the ideal of R generated by the entries of PG is contained in the ideal of R generated by the entries of G. Moreover, if P is invertible over R, then the two ideals coincide.

10. Let J be an ideal in a commutative ring R which is generated as an ideal by a finite number of nilpotent elements. Prove that J is a nilpotent ideal. (Hint: Let J be generated by elements $a_1, a_2, \ldots, a_k$ with indexes of nilpotency $n_1, n_2, \ldots, n_k$. Let

$$N = \max\{n_1, n_2, \ldots, n_k\}$$

Prove that $J^{kN} = 0$.)

11. If A is an $m \times n$ matrix,

$$1 \leq p \leq \min(m,n) \text{ and } 1 \leq i_1 < i_2 < \cdots < i_p,$$
$$1 \leq j_1 < j_2 < \cdots < j_p$$

then the corresponding $p \times p$ minor of A is denoted by

$$A\begin{pmatrix} i_1 & i_2 & \cdots & i_p \\ j_1 & j_2 & \cdots & j_p \end{pmatrix} = \det[a_{i_k j_k}]_{k=1}^{p}$$

If $m \leq n$ and $C = AB$ with A $m \times n$ and B $n \times m$, the Binet-Cauchy theorem (see [21]) says any $p \times p$ minor of C is obtained as follows

$$C\begin{pmatrix} i_1 & i_2 & \cdots & i_p \\ j_1 & j_2 & \cdots & j_p \end{pmatrix} = \sum A\begin{pmatrix} i_1 & i_2 & \cdots & i_p \\ k_1 & k_2 & \cdots & k_p \end{pmatrix} B\begin{pmatrix} k_1 & k_2 & \cdots & k_p \\ j_1 & j_2 & \cdots & j_p \end{pmatrix}$$

where the summation extends over all distinct sets of indices

$$1 \leq k_1 < k_2 < \cdots < k_p \leq n$$

Use the Binet-Cauchy theorem to prove that if P is an idempotent matrix, then $(I_k(P))^2 = I_k(P)$, where $I_k(P)$ is the ideal generated by the $k \times k$ minors of P.

12. Let C be a free module which has rank one as a projective module. Prove that C is free on one generator.

13. Let P be a finitely generated projective module over a commutative ring R. We say that P has *rank* t if the (R/M)-vector space P/MP is t-dimensional for each maximal ideal M of R. Prove: If E is a stably free R-module with $E \oplus R^m = R^n$, then E has rank $n - m$.

14. Suppose the commutative ring R has the following property: If E is a stably free R-module, then $E = Q \oplus \langle u \rangle$, where $\langle u \rangle$ is a rank one free summand. (If, say, $E \subseteq R^t$, then $u \in R^t$ is unimodular). Prove that stably free R-modules are free.

15. For any commutative ring R, call $b \in R$ *irreducible* if $b = xy$, $x,y \in R$, implies that either x or y is invertible. Prove that if b and c are both irreducible and $b \neq xc$ for an invertible x, then R is the only principal ideal containing (b,c).

16. Let $R = Z[-5]$. Prove that ± 1 are the only units in R and that 3 and $\sqrt{-5} - 2$ are both irreducible in Z[-5]. (See the previous exercise.) Then prove that $(3, \sqrt{-5} - 2)$ is not principal.

17. Let $I = (3, \sqrt{-5} - 2)$ in the ring $Z[\sqrt{-5}]$. Verify that

$$-2 + \sqrt{-5} = -3^2 - 3(-2 + \sqrt{-5}) - (-2 + \sqrt{-5})^2$$

and

$$9 = -(2 + \sqrt{-5})(-2 + \sqrt{-5})$$

Conclude that I^2 is a principal ideal with generator $\sqrt{-5} - 2$.

18. If (F,G) is systems equivalent to $(\tilde{F},\tilde{G})$, then

$$\left(\begin{bmatrix} F & 0 \\ 0 & 0 \end{bmatrix}, \begin{bmatrix} G & 0 \\ 0 & I_r \end{bmatrix}\right)$$

is systems equivalent to

$$\left(\begin{bmatrix} \tilde{F} & 0 \\ 0 & 0 \end{bmatrix}, \begin{bmatrix} \tilde{G} & 0 \\ 0 & I_r \end{bmatrix}\right)$$

19. The equilibrium states (= constant solutions) of the real 2-dimensional system

$$x' = \begin{bmatrix} 0 & 0 \\ 0 & 1 \end{bmatrix} x$$

are of the form

$$x = \begin{bmatrix} \lambda \\ 0 \end{bmatrix}$$

where λ is any real number.

a. Linearize the following 2-dimensional system about the above equilibrium states.

$$x_1' = u$$

$$x_2' = x_2 + v$$

$$y_1 = x_1 x_2 + x_1$$

$$y_2 = x_1^2 x_2 - x_2$$

(Expanding the right hand sides of the above equations in Taylor series about the points $x_1 = \lambda$, $x_2 = 0$, $u = 0$, $v = 0$ and keeping only the linear terms, you should get a 2-dimensional linear system (H,F,G), where

$$H = \begin{bmatrix} 1 & \lambda \\ 0 & \lambda^2 - 1 \end{bmatrix}, \quad F = \begin{bmatrix} 0 & 0 \\ 0 & 1 \end{bmatrix}, \quad \text{and } G = \begin{bmatrix} 1 & 0 \\ 0 & 1 \end{bmatrix}$$

b. Show that the dual $(C,A,B) = (G^T,F^T,H^T)$ of the system (H,F,G) is reachable over the ring $\mathbb{R}[\lambda]$ of polynomials in one real variable λ.

c. Explicitly calculate matrices K and L as in the proof of Theorem 3.24 ((2) $\Longrightarrow$ (3)) and verify that the system

$$\left(\begin{bmatrix} A & 0 \\ 0 & 0 \end{bmatrix} + \begin{bmatrix} B & 0 \\ 0 & I_4 \end{bmatrix} K, \begin{bmatrix} B & 0 \\ 0 & I_4 \end{bmatrix} L\right)$$

satisfies the hypotheses of Lemma 3.23.

d. Let

$$\tilde{A} = \left[\begin{array}{c|c} A & 0 \\ \hline 0 & 0 \end{array}\right], \quad \tilde{B} = \left[\begin{array}{c|c} B & 0 \\ \hline 0 & I_3 \end{array}\right]$$

$$K = \left[\begin{array}{cc|ccc} 0 & 0 & 1 & \lambda & -1 \\ 0 & 0 & 0 & -1 & 0 \\ \hline 0 & 0 & 0 & 1 & 0 \\ 0 & 0 & 0 & 0 & 1 \\ 0 & 0 & 0 & 0 & 0 \end{array}\right] \quad \text{and} \quad e_5 = \begin{bmatrix} 0 \\ 0 \\ 0 \\ 0 \\ 1 \end{bmatrix}$$

Verify that $\tilde{A} + \tilde{B}K$ has cyclic vector $e_5 = \tilde{B}e_5$. Hence, using Theorem 3.3, conclude that the system $(\tilde{A},\tilde{B})$ is coefficient assignable and that the system (A,B) is dynamically coefficient assignable with $r = 3 < 4 = n^2$. (This also shows that (A,B) is restrictedly dynamically coefficient assignable with $s = 3$.)

20. Explain why the fact that (F,G) is systems equivalent to $(F + GK,G)$ if and only if

$$\left(\begin{bmatrix} F & 0 \\ 0 & 0 \end{bmatrix}, \begin{bmatrix} G & 0 \\ 0 & I \end{bmatrix}\right)$$

is systems equivalent to

$$\left(\begin{bmatrix} F + GK & 0 \\ 0 & 0 \end{bmatrix}, \begin{bmatrix} G & 0 \\ 0 & I_n \end{bmatrix}\right)$$

is no help in determining the relationship between the integers r and s in the two definitions of dynamic coefficient assignability.

21. Let R be the ring of entire functions defined on the complex plane. Prove that R does not satisfy the hypothesis of Theorem 3.27 by showing that if g is any nonconstant entire function and

$$A = \begin{bmatrix} 0 & g \\ g & 0 \end{bmatrix}$$

then there does not exist an entire function w such that all the eigenvalues of $A(\lambda) + w(\lambda)I$ have positive real parts for all complex λ.

22. Let $R = \mathbf{R}[\lambda]$ be the ring of polynomials in one variable defined over the reals. Let (a,b) be the system with n = 1, m = 1 given by $a(\lambda) = 1$, $b(\lambda) = 1 + \lambda^2$. Prove that this system is reachable for every λ, but that there does not exist a $k(\lambda) \in R$ with $-1 < a(\lambda) + b(\lambda)k(\lambda) < 1$ for all $\lambda \in \mathbf{R}$. (This is related to the stabilization problem for discrete time systems.)

23. Let R be a noetherian ring that is equal to its own total quotient ring (see Section 1.5 for the definition of total quotient ring). Prove that R has only a finite number of maximal ideals (Hint: Use Theorem 1.8 and examine the proof of Theorem 1.7.)

NOTES AND REMARKS

Theorem 3.1 is from [59, Proposition 4.3]. Theorem 3.2 combines Proposition 4.3, Proposition 4.4 and Theorem 4.5 of [33]. The fact that (2) implies (5) in Theorem 3.2 is in [31, p. 49]. Theorem 3.3 is from [28]. Lemma 3.4 is called Heymann's Lemma, see [28]. Our proof of Lemma 3.4 is from [10, p. 11]]. Lemma 3.5, Lemma 3.6, and Theorem 3.7 are from [59, Theorem 4.5 and its proof]. The fact that fields have the pole assignability property was proved in [65] and [47, p. 320]. Theorem 3.8 is from [59, Lemma 4.6]. The fact that neither **R**[X] nor Z has the feedback cyclization property is from [59, p. 23].

Theorem 3.9 has evolved over several years. Our proof is from [8] with ideas from [7] and [10]. Theorem 3.10 appears in [8] with ideas from [45] and [19]. In fact, the lemma in the proof of 3.10 is from [19]. The idea for Theorem 3.11 occurs in [10, Example 3.9] and [63]. Theorem 3.12 is the sharpening of [10, Proposition 3.7] given in [9]. Theorem 3.13 is from [10, p. 119]. The fact that principal ideal domains, in particular **R**[X], have the pole assignability property was proved earlier in [43]. Theorems 3.14 and 3.15 are stated without proof in [37, p. 80]. Professor Kaplansky provided us the proofs. Theorem 3.19 with the ring of real analytic functions replaced by the ring of entire functions is in the literature. Specifically, it is proved in [26] that the ring of entire functions is infinite dimensional; also see [44, p. 540]. Our proof of Theorem 3.17 is based on the proof that the ring of entire functions is infinite-dimensional given in [40]. Theorem 3.18 is from [10, Example 3.8]. Proofs that the ring of real analytic function is a Bézout domain are in [10, p. 120] and [58, p. 335]. It was known that the ring of entire functions is completely integrally closed; see [22, Exercise 16, p. 147].

Theorem 3.20 originated in our attempts to understand Theorems 3.21 and 3.22. The proof of Theorem 3.20 is from [7]. Theorem 3.21 was proved in [63]; the case when k is algebraically closed was proved slightly earlier in [62]. Theorem 3.22 is from [10, Example 3.10].

The proof of Theorem 3.24 follows Sontag [61], who attributes this proof to P. P. Khargonekar (cf. [35]). Lemma 3.23 is classical for fields. The idea of using the transition matrix Φ in the proof of Lemma 3.23 was taken from [56].

Theorem 3.27 is an axiomatized version of [60]. Stabilizability of systems depending on parameters was studied earlier in [11] and [12]. In particular, it is proved in [12] that pointwise reachability implies stabilizability for systems defined over commutative Frechet algebras.

Theorems 3.28 and 3.29 are from [14, Theorems 3.7 and 3.8].

Exercise 15 is due to Daniel Katz. The example in Exercises 16 and 17 is from [10]. Parts (a), (b), and (c) of Exercise 19 are from [61], as is Exercise 22.

4

Realization Theory

It is easy to state the problem with which this chapter is concerned.

Let R be a commutative ring with $\{A_i\}_{i=1}^{\infty}$ a collection of $p \times m$ matrices over R. What are necessary and sufficient conditions on the collection $\{A_i\}$ in order that there exist a $p \times n$ matrix H, an $n \times n$ matrix F, and an $n \times m$ matrix G such that $A_i = HF^{i-1}G$ for $i \geq 1$? In case such matrices H, F, and G exist, we say that the system (H,F,G) is a *realization* of $\{A_i\}$. Notice that each A_i, being a $p \times m$ matrix, can be considered as an R-homomorphism from $R^m \rightarrow R^p$. Moreover, the sizes of H, F^{i-1}, and G mean that they give rise to the following diagram of free modules and R-homomorphisms:

$$R^m \xrightarrow{G} R^n \xrightarrow{F^{i-1}} R^n \xrightarrow{H} R^p \quad \text{for } i \geq 1$$

Therefore, the condition that $A_i = HF^{i-1}G$ says precisely that we have algebraically factored A_i through R^n—via H, F^{i-1}, and G. In this case we say that $\{A_i\}$ is n-*realizable* or simply *realizable.*

We recall from Chapter 0 that if

$$\frac{dx}{dt} = Fx(t) + Gu(t)$$

$$y(t) = Hu(t)$$

is a linear system, then assuming that $x(0) = 0$, the Laplace transform of $y(t)$ is

$$Y(s) = \left\{ \sum_{i=1}^{\infty} HF^{i-1}Gs^{-i} \right\} U(s)$$

The function $Y(s)/U(s)$ is called the *transfer function* of the linear system. What we are seeking in this chapter is a characterization of when a given function of the form

$$\sum_{i=1}^{\infty} A_i s^{-i}$$

is actually realizable as the transfer function of a linear system. We shall discuss this problem at length and we shall find that it leads us to a great deal of commutative algebra. In the end we will have answered our question in several different ways and for many different kinds of rings. For certain types of rings, we will not only answer the question, but will also give an algorithm for obtaining the matrices H, F, and G. This is a most useful thing in practice.

As ever, our point of view is that the answer to the question is known for fields, even classical (= twenty years old or more), and that we seek the answer for a broader class of rings.

Throughout the chapter we shall call the collection of matrices $\{A_i\}$ an i/o *map* for "input-output map" and we shall retain the convention that each A_i is a $p \times m$ matrix.

4.1. HANKEL MATRICES, RECURRENCE, AND THE BASIC REALIZATION THEORY

Given an i/o map $f = \{A_i\}$, we can associate with it in a natural way its *behavior* or *Hankel matrix* $B(f)$ as follows:

$$B(f) = \begin{bmatrix} A_1 & A_2 & A_3 & \cdots \\ A_2 & A_3 & A_4 & \cdots \\ A_3 & A_4 & A_5 & \cdots \\ \vdots & \vdots & \vdots & \end{bmatrix}$$

This matrix is a useful device in the study of realization theory. In essence, the "finiteness" of B(f) is equivalent to the realizability of f. This idea will, hopefully, become clear over the course of this chapter. In fact, our first result is in this direction. We begin by fixing some notation and terminology.

Let $f = \{A_i\}_{i=1}^{\infty}$ be an i/o map over a commutative ring R and let B(f) denote the Hankel matrix of f. We shall write X_f for the R-module generated by the columns of B(f). We shall say that f is t-*recurrent* if there exist elements $r_1, r_2, \ldots, r_{t-1} \in R$ so that

$$A_{t+k} = \sum_{i=1}^{t-1} r_i A_{i+k} \quad \text{for } k \geqslant 0$$

Obviously, recurrence of f is a form of finiteness of f. This is related to realizability of f by Theorem 4.1.

THEOREM 4.1. Let $f = \{A_i\}_{i=1}^{\infty}$ be an i/o map over a commutative ring R. Then

1. If the map f is n-realizable, then it is (n + 1)-recurrent.

2. If f is n-recurrent, then it is [(n - 1)m]-realizable, and X_f is generated by at most (n - 1)m elementary columns. In particular, X_f is a finitely generated R-module.

Proof. (1): Suppose that f is n-realizable and let (H, F,G) be a realization with F an n × n matrix. By the Cayley-Hamilton theorem, there exist elements $r_0, r_1, \ldots, r_{n-1} \in R$ so that $F^n + r_{n-1}F^{n-1} + \cdots + r_1F + r_0 = 0$ — that is,

$$F^n = \sum_{i=0}^{n-1} (-r_i)F^i$$

Therefore,

$$F^{n+k} = \sum_{i=0}^{n-1} (-r_i)F^{i+k} \quad \text{for } k \geqslant 0$$

and so

$$A_{n+1+k} = HF^{n+k}G = \sum_{i=0}^{n-1} (-r_i)HF^{i+k}G = \sum_{i=0}^{n-1} (-r_i)A_{i+k+1} \quad \text{for } k \geqslant 0$$

It follows that f is (n + 1)-recurrent.

(2): Suppose that f is n-recurrent and that the notation is

$$A_{n+k} = \sum_{i=1}^{n-1} r_i A_{i+k} \quad \text{for } k \geqslant 0$$

We claim that the system (H,F,G) as given below is a realization of f of dimension (n − 1)m.

$$H = [A_1, A_2, \ldots, A_{n-1}],\ F = \begin{bmatrix} 0 & 0 & 0 & \ldots & r_1 I \\ I & 0 & 0 & \ldots & r_2 I \\ 0 & I & 0 & \ldots & r_3 I \\ \vdots & \vdots & I & & \\ & & \vdots & & \\ 0 & 0 & 0 & \ldots\ I & r_{n-1} I \end{bmatrix},\ G = \begin{bmatrix} I \\ 0 \\ 0 \\ \vdots \\ 0 \end{bmatrix}$$

The characteristic polynomial of F is $X^{n-1} - r_{n-1}X^{n-1} - \cdots - r_2X - r_1$ and so

$$F^{n-1} = \sum_{i=1}^{n-1} r_i F^{i-1}$$

from which it follows that

$$F^{n+k-1} = \sum_{i=1}^{n-1} r_i F^{i-1+k} \quad \text{for } k \geqslant 0$$

Therefore,

$$HF^{n+k-1}G = \sum_{i=1}^{n-1} r_i HF^{i-1+k}G$$

But we also have that $A_{n+k} = \sum_{i=1}^{n-1} r_i A_{i+k}$ and therefore, if we knew that, for $1 \leqslant j \leqslant n-1$, $HF^{j-1}G = A_j$, we would have that $HF^{i-1}G = A_i$ for $i \geqslant 1$.

To prove that $HF^{j-1}G = A_j$ for $1 \leqslant j \leqslant n-1$, we begin by computing the first column of F^j. We claim that it has I in the $(j+1)$st block row and zeros elsewhere. This is true for $j = 1$ and assume it for $j - 1$. Then

$$F^j = F \cdot F^{j-1} = \begin{array}{c} \\ \\ \\ \\ j+1 \\ \\ \end{array} \begin{bmatrix} 0 & 0 & \cdots & \overset{j}{0} & \cdots \\ I & & & \vdots & \\ \vdots & & & \vdots & \\ 0 & & \cdots & I & \cdots \\ \vdots & & & & \end{bmatrix} \begin{array}{c} \\ \\ j \\ \\ \\ \end{array} \begin{bmatrix} 0 & \cdots \\ \vdots & \\ I & \cdots \\ \vdots & \\ 0 & \end{bmatrix} = \begin{array}{c} \\ \\ j+1 \\ \\ \\ \end{array} \begin{bmatrix} 0 & \cdots \\ \vdots & \\ I & \\ \vdots & \\ 0 & \cdots \end{bmatrix}$$

and moreover this works for $1 \leqslant j \leqslant n-1$. Now, look at $F^{j-1}G$ Since G is nonzero only in the first block row, $F^{j-1}G$ picks off the first block column of F^{j-1} which, as we saw above, is zero save for the I in block row j. Thus, $H \cdot F^{j-1}G$ is A_j as desired. Since I is $m \times m$, the realization is of dimension $(n-1)m$.

Also, from the relation $A_{n+k} = \sum_{i=1}^{n-1} r_i A_{i+k}$ for $k \geqslant 0$, we see that the jth elementary column of the $(n+k)$th block colu of $B(f)$ is a linear combination of the jth elementary columns of block columns $k+1,\ldots,k+n-1$. This holds for $k \geqslant 0$ and, since it does hold for $k = 0$, we see that the $m \cdot (n-1)$ elementary columns from blocks 1 through $n-1$ generate X_f.

Our next goal is to prove that, if X_f is a finitely generated R-module, then f is recurrent and, hence, realizable.

Accomplishing this will require some effort on our parts. Indeed, we come to the essential ideas involved in obtaining a realization of an i/o map f over a ring R. The first trick is to extend our definition of system so as to allow for more general "state spaces." Thus, temporarily, we shall take a system to be a quadruple (X,H,F,G) where X is an R-module, F an R-endomorphism of X, G an R-homomorphism from R^m into X, and H an R-homomorphism from X into R^p. Then we shall say that (X,H,F,G) is a *realization* of $f = \{A_i\}$ over R if and only if $A_i = H \circ F^{i-1} \circ G$ for $i \geq 1$. We said "temporarily," but, in fact, we shall retain this notation for some time. If the system is written as (H,F,G), it is understood that the state module is R^n. The reader need not be put off by this "pseudo-realization" for we shall eventually prove that in case X is finitely generated such pseudo-realizations can always be lifted to realizations where the state module is R^n. For the time being, this new definition of realization is a very handy notion with which to work.

The task before us is to obtain a realization of the i/o map $f = \{A_i\}$. We begin by forming the Hankel matrix B(f) and the column module X_f. We next define the R-homomorphisms which, together with the R-module X_f, will be our realization of f. Suppose as usual that each A_i is a $p \times m$ matrix.

Define $\beta : R^m \longrightarrow X_f$ by $\beta(\varepsilon_i) = C_i$, where ε_i is the ith elementary basis element of R^m and C_i is the ith elementary column from the first block of B(f). Since β is defined on a basis of the free module R^m, β extends uniquely to an R-homomorphism from R^m into X_f. We call β the "blow-up" because it blows up R^m into the first block column of B(f).

Define $\sigma_p : X_f \longrightarrow X_f$ as follows. Let R^∞ denote the countably infinite rank, free R-module of long columns of elements of R. Define $\sigma_p^* : R^\infty \longrightarrow R^\infty$ by $\sigma_p^*(\varepsilon_i^\infty) = \varepsilon_{i-p}^\infty$, where ε_i^∞ is the ith elementary basis element of R^∞ and our convention is that $\varepsilon_j^\infty = 0$ if $j \leq 0$. By the definition of the Hankel matrix, $\sigma_p^*(C_i) = C_{i+m}$ for any elementary column C_i of B(f). Since the columns generate X_f, it follows that $\sigma_p^*(X_f) \subseteq X_f$ and hence σ_p^* restricts to an R-homomorphism $\sigma_p : X_f \longrightarrow X_f$ called the "p-shift" because, for example, it shifts block column i to block column i + 1.

Define $\gamma_p : X_f \longrightarrow R^p$ as follows. With R^∞ as above, define $\gamma_p^* : R^\infty \longrightarrow R^p$ by $\gamma_p^*(\varepsilon_i^\infty) = \varepsilon_i$ for $1 \leq i \leq p$ and $\gamma_p^*(\varepsilon_i^\infty) = 0$ for $i > p$. Then γ_p^* restricts to an R-homomorphism $\gamma : X_f \longrightarrow R^p$ called the "chop" because it chops off the top of a column of B(f).

We are now able to give the basic theorem in realization theory.

THEOREM 4.2. Let $f = \{A_i\}$ be an i/o map over a commutative ring R with each A_i a $p \times m$ matrix. Let B(f) be the Hankel matrix of f with X_f the R-module generated by the columns. Let S_1 be the standard basis of R^m and S_2 be the standard basis of R^p. Then

1. ${}_{S_1}[\gamma_p \circ \sigma_p^{i-1} \circ \beta]_{S_2} = A_i$ for $i \geq 1$. In words, the matrix of the homomorphism $\gamma_p \circ \sigma_p^{i-1} \circ \beta$ with respect to the bases S_1 and S_2 is A_i.

2. $(X_f, \gamma_\rho, \sigma_p, \beta)$ is a realization of f.

Proof. By looking at B(f) and the definitions of the homomorphisms β, σ_p, and γ_p, this becomes quite obvious. Indeed, the work involved in the theorem was done when we defined the homomorphisms.

Customarily, we seek realizations where the state module is finitely generated. Conditions for this are given in Theorem 4.4. First we need a Cayley-Hamilton type theorem for finitely generated modules over a commutative ring.

THEOREM 4.3. Let R be a commutative ring with X an R-module generated by n elements. If F is an R-endomorphism of X, then F satisfies a monic polynomial of degree n—that is $F^n + \sum_{i=0}^{n-1} a_i F^i$ is the zero homomorphism for some $a_0, \ldots, a_{n-1} \in R$.

Proof. Suppose that $x_1, \ldots, x_n$ generate X and, for $1 \leqslant i \leqslant n$, write $Fx_i = \sum_{j=1}^{n} r_{ij} x_j$ with $r_{ij} \in R$. This leads to the following situation.

$$\begin{array}{ccc} (r_{11}I - F)x_1 + \cdots + r_{1n}Ix_n & = & 0 \\ \vdots & & \vdots \\ r_{n1}Ix_1 + \cdots + (r_{nn}I - F)x_n & = & 0 \end{array}$$

where I denotes the identity homomorphism on X. We can regard this as saying that $x_1, \ldots, x_n$ is a solution to a certain set of linear homogeneous equations where the coefficients are in the ring R[F], the commutative ring of all polynomials in F with coefficients from R. By Cramer's Rule, if $d = \det([r_{ij} - F\delta_{ij}I])$, where δ_{ij} denotes the Kronecker delta, then $dx_i = 0$ for $1 \leqslant i \leqslant n$. Since d annihilates the generators of X, it follows that d annihilates X. But, d is a monic polynomial in F and this completes the proof.

We now prove that, if X_f is finitely generated, then f is realizable. Recall that here "realizable" allows homomorphisms in lieu of matrices.

THEOREM 4.4. The i/o map f is realizable with finitely generated state module if and only if X_f is finitely generated.

Proof. (<==): This follows directly from Theorem 4.2 since X_f is the state module of the realization $(X_f, \gamma_p, \sigma_p, \beta)$.

(==>): If (X,H,F,G) is a realization of f with X finitely generated, say by n elements, then by Theorem 4.3, there exist elements $r_0, r_1, \ldots, r_{n-1} \in R$ so that $F^n + r_{n-1}F^{n-1} + \cdots + r_1F + r_0 = 0$. Therefore,

$$F^{n+k} = \sum_{i=0}^{n-1} (-r_i)F^{i+k} \text{ for } k \geq 0$$

Since (X,H,F,G) is a realization of f, it follows that f is recurrent and, from Theorem 4.1, part (2), that X_f is finitely generated.

4.2. HANKEL MATRICES OVER FIELDS AND SILVERMAN'S FORMULAS

Given an i/o map f over a ring R, we saw in the last section how useful the Hankel matrix B(f) is in studying the realization problem. It is natural to expect that the Hankel matrix would yield even more information under stronger assumptions on the ring and that is indeed the case. In this section we shall be studying Hankel matrices over fields, but we begin with some results about arbitrary countably infinite by countably infinite matrices over a field.

Let M be a countably infinite by countably infinite matrix over a field L. By the *rank of* M we mean the largest positive integer n such that there exists an $n \times n$ submatrix M_1 of M such that $\det(M_1) \neq 0$, but all larger submatrices have zero determinants. By the *column space* of M we mean the L-vector space spanned by the columns of M, and by the *row space* of M we mean the L-vector space spanned by the rows of M.

THEOREM 4.5. Let M be a countably infinite by countably infinite matrix over a field L. The following are equivalent.

1. The rank of M equals $n < \infty$.
2. The dimension of the column space of M equals $n < \infty$.
3. The dimension of the row space of M equals $n < \infty$.

Proof. (1) $\Longrightarrow$ (2): We shall prove the following: Let M^* be any $n \times n$ submatrix of M such that $\det(M^*) \neq 0$. If $C_1, \ldots, C_n$ are the columns of M corresponding to the columns of M^*, then $\{C_1, \ldots, C_n\}$ is a basis for the column space of M.

If $r_1C_1 + r_2C_2 + \cdots + r_nC_n = 0$ for $r_i \in L$, the same relation holds among the columns of M^*. Since $\det(M^*) \neq 0$, the columns of M^* are linearly independent and so $r_i = 0$ for $1 \leqslant i \leqslant n$. Therefore, $\{C_1, \ldots, C_n\}$ is a linearly independent set.

Suppose that C is any column of M distinct from $C_1, \ldots, C_n$. Let C^* denote the restriction of C to those n rows involved in M^* with $C_1^*, \ldots, C_n^*$ the columns of M^*. Then $\{C_1^*, \ldots, C_n^*, C^*\}$ is a linearly dependent set and hence there exist elements $r_1, \ldots, r_n \in L$ so that $C^* = r_1C_1^* + \cdots + r_nC_n^*$. If $C \neq r_1C_1 + \cdots + r_nC_n$, then there must exist some row of M, say row j, different from those of M^* such that the entry of C in that row is unequal to $r_1C_{1,j} + \cdots + r_nC_{n,j}$. Now look at the new matrix

$$A = \begin{bmatrix} M^* & C^* \\ - & C_j \end{bmatrix}$$

where $-$ denotes that portion of row j corresponding to M^*. By subtracting r_i times the ith column of A from

$$\begin{bmatrix} C^* \\ C_j \end{bmatrix}$$

we are left with the final column of A having only zeros except in the lower right hand corner where we have, say, $x \neq 0$. Then $\det(A) = \det(M^*) \cdot x \neq 0$. This yields the contradiction that $\operatorname{rank}(M) \geqslant n + 1$. Therefore, $C = r_1C_1 + \cdots + r_nC_n$ and $\{C_1, \ldots, C_n\}$ spans the column space of M.

It is clear that the same argument works on the row space of M. Thus, condition (1) also implies condition (3). Moreover, by symmetry, we have only to prove that condition (2) implies condition (1).

(2) ⟹ (1): Suppose that the dimension of the column space of M equals n. If rank(M) $< \infty$, then it must be n by the first part of the theorem. But if rank(M) $= \infty$, then by definition M contains a submatrix N of size $(n + 1) \times (n + 1)$ with a nonzero determinant. By the proof of (1) implies (2), the columns of M corresponding to the columns of N are linearly independent, contradicting the fact that the column space of M has dimension n. This completes the proof.

Even finer results are possible in the case of a Hankel matrix.

THEOREM 4.6. Suppose that L is a field and that $f = \{A_i\}$ is an i/o map over L with each A_i a $p \times m$ matrix. If rank(B(f)) $= n < \infty$, then the first n block columns of B(f) contain a basis for X_f, the column space of B(f). In particular, in the "single input" case when $m = 1$, the first n columns of B(f) are linearly independent. Moreover, let $C_1,\ldots,C_n$ be n linearly independent long columns of B(f) occurring in the first n block columns of B(f). Assume the notation is such that C_i is to the left of C_{i+1}. Then the n short columns $\hat{C}_1,\ldots,\hat{C}_n$ formed by the parts of $C_1,\ldots,C_n$, respectively, lying in the first n block rows are also linearly independent.

Thus, in the scalar case, $m = p = 1$, if rank(B(f)) $= n$, then the determinant of the $n \times n$ matrix in the upper left corner is nonzero.

Proof. Denote by $\bar{A}_i$ the block column headed by the matrix A_i. We first prove the single input case $m = 1$. It suffices to prove the following: If column $\bar{A}_t$ is linearly dependent upon columns $\bar{A}_1,\ldots,\bar{A}_{t-1}$, then for $j \geq 0$, column $\bar{A}_{t+j}$ is linearly dependent upon columns $\bar{A}_1,\ldots,\bar{A}_{t-1}$. For suppose that we have established the claim. Since rank(B(f)) = n < ∞, the column space X_f has dimension n by Theorem 2.5. It follows that column $\bar{A}_1$ is a linearly independent column and if columns $\bar{A}_1,\dots,\bar{A}_{t-1}$ are linearly independent, but columns $\bar{A}_1,\dots,\bar{A}_t$ are linearly dependent, then by the claim, the dimension of X_f equals t - 1.

As for the claim itself, assume we know that for $0 \leq j \leq u - 1$, $\bar{A}_{t+j}$ is linearly dependent upon $\bar{A}_1,\dots,\bar{A}_{t-1}$. If $\bar{A}_{t+(u-1)} = r_1\bar{A}_1 + \cdots + r_{t-1}\bar{A}_{t-1}$, then

$$\begin{aligned}\bar{A}_{t+u} &= \sigma_p(\bar{A}_{t+(u-1)}) = r_1\sigma_p\bar{A}_1 + \cdots + r_{t-1}\sigma_p\bar{A}_{t-1}\\ &= r_1\bar{A}_2 + \cdots + r_{t-1}\bar{A}_t\end{aligned}$$

Since $\bar{A}_t$ is linearly dependent upon $\bar{A}_1,\dots,\bar{A}_{t-1}$, it follows that $\bar{A}_{t+u}$ is as well. This completes the proof in case m = 1.

Concerning the general case, first observe the following. For $1 \leq j \leq m$, denote by $\bar{A}_i^{(j)}$ the jth elementary column from block column $\bar{A}_i$. The submatrix $[\bar{A}_1^{(j)},\bar{A}_2^{(j)},\dots]$ of B(f) is a single input Hankel matrix and, since rank(B(f)) = n,

$$\operatorname{rank}([\bar{A}_1^{(j)}, \bar{A}_2^{(j)},\dots]) \leq n \text{ for each } j$$

So, consider the matrix $[\bar{A}_1^{(1)}, \bar{A}_2^{(1)},\dots]$. If this matrix has rank n, then $\{\bar{A}_1^{(1)},\dots,\bar{A}_n^{(1)}\}$ is a basis for X_f, by the single input case. Otherwise, let t + 1 be the least integer such that $\bar{A}_1^{(1)},\dots,\bar{A}_{t+1}^{(1)}$ are linearly dependent upon the linearly independent columns $\bar{A}_1^{(1)},\dots,\bar{A}_t^{(1)}$.

Next consider $[\bar{A}_1^{(2)},\bar{A}_2^{(2)},\dots]$. If $\bar{A}_k^{(j)}$ is linearly dependent upon $\bar{A}_1^{(2)},\dots,\bar{A}_{k-1}^{(2)}$ and $\bar{A}_1^{(1)},\dots,\bar{A}_t^{(1)}$, then all subsequent second columns are also. To see this, merely apply the shift operator as we did in the first part of the proof.

Continuing in this fashion for 3,...,m, we eventually reach a basis for the space X_f. Moreover, since rank($[\bar{A}_1^{(j)}$,

$\bar{A}_2^{(j)},\ldots]) \leq n$ for $1 \leq j \leq n$, in no case do we ever have to go past $\bar{A}_n^{(j)}$ before reaching a dependency relation. Thus, the first n block columns of B(f) contain a basis for X_f.

As for the moreover assertion, let $\hat{C}_k$ be the first short column that is linearly dependent on the earlier columns $\hat{C}_1, \ldots, \hat{C}_{k-1}$. Note that k might equal 1. Write $\hat{C}_k = r_1\hat{C}_1 + \cdots + r_{k-1}\hat{C}_{k-1}$ with $r_1, \ldots, r_{k-1} \in L$. Consider the long vector

$$\alpha_1 = C_k - \sum_{i=1}^{k-1} r_i C_i$$

which belongs to X_f. The first n block rows of this vector are zero, but the vector itself is nonzero since $C_1, \ldots, C_n$ are linearly independent. Let $j > n$ be the first block row that contains a nonzero element. Apply the shift operator σ_p to the vector α_1 to obtain

$$\alpha_2 = \sigma_p C_k - \sum_{i=1}^{k-1} r_i \sigma_p C_i$$

Then α_2 also belongs to X_f and is linearly independent of α_1 because the first (j - 2) block rows of α_2 are zero and its (j - 1)st block row is nonzero. To illustrate the process one step further, set

$$\alpha_3 = \sigma_p \alpha_2 = \sigma_p^2 C_k - \sum_{i=1}^{k-1} r_i \sigma_p^2 C_i$$

The first (j - 3) block rows of α_3 are zero, while the (j - 2)nd block row is the jth block row of α_1 and this is nonzero. Hence, $\{\alpha_1, \alpha_2, \alpha_3\}$ is a linearly independent set. We may repeat the process a total of (j - 1) times to get j linearly independent vectors in the column space of B(f). But $j > n$ and rank (B(f)) = n are contradictory statements and hence no such integer k exists. Note that the process even works if $k = 1$ for now $\alpha_1 = C_1$. This completes the proof of Theorem 4.6.

REMARK. Of course, Theorem 4.6 still leaves open the problem of how to actually calculate rank(B(f)); indeed, the problem of determining whether or not the rank is finite is unresolved. On the other hand, the process given in the proof is an optimal strategy <u>in some sense</u> for determining whether or not the rank is finite.

In the notation of Theorem 4.2, notice that if X_f is a finitely generated R-module with basis B, then

$$({}_{B}[\gamma_p]_{S_2},\ {}_{B}[\sigma_p]_{B},\ {}_{S_1}[\beta]_{B})$$

is a matrix realization of $\{A_i\}$. Since any module over a field is free, we have the following theorem.

THEOREM 4.7. Let $f = \{A_i\}$ be an i/o map over a field L and denote by B(f) the Hankel matrix of f. The following are equivalent.

1. The map f is realizable of minimum order n.
2. The rank of B(f) is finite of order n.
3. X_f is an n-dimensional vector space over L.

Moreover, in the event one, and hence all, of these equivalent conditions are satisfied, the first n block columns of B(f) contain a basis for X_f.

Proof. Statement (2) is equivalent to (3) by Theorem 4.5, and Theorem 2.2 shows that, assuming (3), f has a realization of order n. Conversely, if f has a realization (H,F,G) of order k, then X_f is finite-dimensional by Theorem 4.1. Then by Theorem 2.5, there is an integer N such that the upper left $N \times N$ block of B(f) has rank equal to the dimension of X_f. But this upper left block of B(f) factors as

$$\begin{bmatrix} H \\ HF \\ \vdots \\ HF^{N-1} \end{bmatrix} \cdot \left[G, FG, F^2G, \ldots, F^{N-1}G\right]$$

The second factor has row rank at most k, hence X_f has dimension at most k. This completes the proof.

Given an i/o map f over a field we would like to explicitly determine a matrix realization of it, if one exists, and although the first n block columns of B(f) contain a basis for the state space X_f, we still need a method for actually finding the matrices themselves. Before giving such a method, we record an immediate corollary to Theorem 4.6 which says, in essence, that if we know that the Hankel matrix has rank n, finite, we can restrict our attention to the finite matrix $B_{n,n}$, the n block row by n block column submatrix in the upper left corner of B(f).

THEOREM 4.8. Let $f = \{A_i\}$ be an i/o map over an integral domain D and let L be the quotient field of D. Assume that $\text{rank}_L(B(f)) = n < \infty$. Let $B_{n,n}$ be the submatrix of B(f) consisting of the first n block rows and block columns. Then $B_{n,n}$ contains an $n \times n$ submatrix having nonzero determinant.

Before stating our next result, we need a great deal of additional notation. Suppose that D is an integral domain with quotient field L and let $f = \{A_i\}$ be an i/o map over D. If X_f is a free D-module of rank n, then the tensor product $X_f \otimes_D L$ is an L-vector space of dimension n. By Theorem 4.8, there exists an $n \times n$ submatrix Φ^* of $B_{n,n}$ such that $\det(\Phi^*) \neq 0$. If $(\Phi^*)^{(i)}$ denotes the ith column of the matrix Φ^*, and if $\Phi^{(i)}$ denotes the corresponding long column of B(f), the set $\mathcal{B}^* = \{(\Phi^*)^{(i)}\}_1^n$ forms an L-basis for L^n and the set $\mathcal{B} = \{\Phi^{(i)}\}_1^n$ forms an L-basis for $X_f \otimes_D L \cong L^n$. Set

$$\Gamma = B_{n,1} \cap \text{(rows of } \Phi^* \text{ extended)}$$

and

$$\Lambda = B_{1,n} \cap \text{(columns of } \Phi^* \text{ extended)}$$

Let $\sigma_p(\Phi^*)$ denote the n × n matrix whose ijth entry is the (i + p)th coordinate of $\Phi^{(j)}$. We can now give the explicit realization theorem for fields. We retain the same notation as in the basic theorem of realization theory, Theorem 4.2.

THEOREM 4.9. Let L be a field with $f = \{A_i\}$ an i/o map over L. Suppose that $\text{rank}(B(f)) = n < \infty$ and that Φ^* is an n × n submatrix of $B_{n,n}$ with $\det(\Phi^*) \neq 0$. Then

1. $\quad {}_{S_1}[\beta]_{B} = {}_{S_1}[\Gamma]_{B*}$

2. $\quad {}_{B}[\sigma_p]_{B} = {}_{B*}[\sigma_p(\Phi^*) \cdot (\Phi^*)^{-1}]_{B*}$

3. $\quad {}_{B}[\gamma_p]_{S_2} = {}_{B*}[\Lambda \cdot (\Phi^*)^{-1}]_{S_2}$

Therefore, $(\Lambda \cdot (\Phi^*)^{-1}, \Phi_p(\Phi^*) \cdot (\Phi^*)^{-1}, \Gamma)$ is a matrix realization of f. These are called "Silverman's Formulas."

Proof. If claims (1)-(3) have been verified, then the mapping $(\gamma_p \circ \sigma_p^j \circ \beta) : L^m \longrightarrow L^p$ will have been matrix factored (through L^n) by the product $\Lambda \cdot (\Phi^*)^{-1} \cdot [\sigma_p(\Phi^*) \cdot (\Phi^*)^{-1}]^j \cdot \Gamma$ and the result will follow from the basic theorem, Theorem 4.2.

(1): For $1 \leqslant j \leqslant m$,

$$\Gamma^{(j)} = \Gamma \cdot \varepsilon_j = \beta(\varepsilon_j) \cap \text{(rows of } \Phi^* \text{ extended)}$$

is, by the proof of Theorem 4.5, the same linear combination of $(\Phi^*)^{(1)}, \ldots, (\Phi^*)^{(n)}$ as $\beta(\varepsilon_j)$ is of $\Phi^{(1)}, \ldots, \Phi^{(n)}$. The result follows. (What we are using here is the following: If C is a long column of B(f) and if

$$C^* = C \cap \{\text{rows of } \Phi^* \text{ extended}\}$$

then

$$C = \sum_{i=1}^{n} d_i \cdot \Phi^{(i)} \text{ if and only if } C^* = \sum_{i=1}^{n} d_i \cdot (\Phi^*)^{(i)}$$

(2): For $1 \leq j \leq n$,

$$[\sigma_p(\Phi^*) \cdot (\Phi^*)^{-1}] \cdot (\Phi^*)^j = \sigma_p(\Phi^*) \cdot \varepsilon_j = [\sigma_p(\Phi^*)]^{(j)}$$
$$= \sigma_p[(\Phi^*)^{(j)}] = \sigma_p(\Phi^{(j)}) \cap \text{ (rows of } \Phi^* \text{ extended)}$$

Just as in (1), it follows that

$$_{B}[\sigma_p]_{B} = {}_{B^*}[\sigma_p(\Phi^*) \cdot (\Phi^*)^{-1}]_{B^*}$$

(3): For $1 \leq j \leq n$,

$$[\Lambda \cdot (\Phi^*)^{-1}] \cdot (\Phi^*)^{(j)} = \Lambda \cdot \varepsilon_j = \Lambda^{(j)} = \gamma_p(\Phi^{(j)})$$

Thus,

$$_{B}[\gamma_p]_{S_2} = {}_{B^*}[\Lambda \cdot (\Phi^*)^{-1}]_{S_2}$$

This completes the proof.

Via Silverman's Formulas we can actually determine a matrix realization of an i/o map f over a field L if one exists. Indeed, once we know <u>by some means</u>, that $\text{rank}_L(B(f)) = n < \infty$, then we look at the (finite) matrix $B_{n,n}$. It has rank n and we can, by any of several methods, find an $n \times n$ submatrix Φ^* of $B_{n,n}$ with nonzero determinant. We then invert Φ^*, find $\sigma_p(\Phi^*)$, Γ, and Λ and write down the realization that Silverman's Formulas yield.

As stated in the beginning, our point of view is that the field case is well known and it is our goal to prove results over more general rings. In particular, we would like to extend the class of rings to which Silverman's Formulas apply.

Given an integral domain D with quotient field L, we shall need to know two things: firstly, that realizability over L implies realizability over D and, secondly, that we can do a certain amount of linear algebra over D. In particular, we shall want to know that finitely generated submodules of free modules are free. This last condition is precisely the Bézout domain condition. (See Theorem 1.20.) We shall devote an entire section to the problem of when realizability over L implies realizability over D, but for now we say only that if D is completely integrally closed, then realizability over L implies realizability over D.

Therefore, our next theorem will be stated for completely integrally closed Bézout domains. Examples of such rings include principal ideal domains, one-dimensional Bézout domains, and the domain of all real analytic functions on an open interval. (See Chapter 1 and Theorem 3.19.)

Our proof will be constructive yielding, as in the field case, an algorithm for finding a realization if one exists. To single out the algorithm as well as to fix notation, we first explicitly write down its steps.

Let D be a completely integrally closed Bézout domain with quotient field L and let $f = \{A_i\}$ be an i/o map over D. Suppose further that $\text{rank}_L(B(f)) = n < \infty$ so that f is realizable over L and hence over D. Then a realization of f over D can be found as follows:

1. Choose an $n \times n$ submatrix Φ of $B_{n,n}$ having nonzero determinant.

2. Let Δ be the $n \times (mn)$ submatrix of $B_{n,n}$ whose rows correspond to the rows of Φ. Let J_1 be the ideal of D generated by the elements of the first row of Δ and suppose that $J_1 = a_1D$. Then a_1 is a D-linear combination of elements of

the first row of Δ. Let u_1^* be the corresponding linear combination of the columns of Δ.

3. Zero the top row of Δ by subtracting from each column a suitable multiple of u_1^*. In this way a matrix Δ_1 is obtained whose top row consists entirely of zeros.

4. Apply the same procedure to Δ_1 to obtain a vector u_2^* and a matrix Δ_2 whose top two rows are zero.

5. Continuing in this fashion, we arrive at a basis $\{u_1^*, u_2^*, \ldots, u_n^*\}$ for the free D-module generated by the columns of Δ.

6. Setting $\Gamma = \Delta \cap B_{n,1}$ and

$$\Lambda = (\text{"long" columns of } \Phi) \cap B_{1,n}$$

we obtain Silverman's realization of f over D:

$$H = \Lambda \cdot \Phi^{-1} \cdot [u_1^*, u_2^*, \ldots, u_n^*]$$

$$F = [u_1^*, u_2^*, \ldots, u_n^*]^{-1} \cdot (\sigma_p(\Phi)) \cdot \Phi^{-1} \cdot [u_1^*, u_2^*, \ldots, u_n^*]$$

and

$$G = [u_1^*, u_2^*, \ldots, u_n^*]^{-1} \cdot \Gamma$$

Now for the theorem itself. We retain the above definitions of Δ and Φ.

THEOREM 4.10. Let D be a completely integrally closed Bézout domain with quotient field L. Let $f = \{A_i\}$ be a $p \times m$ i/o map over D with $\text{rank}_L(B(f)) = n < \infty$. Then:

1. There exists a basis $\mathcal{B}_1^* = \{u_1^*, \ldots, u_n^*\}$ for the free D-module X_f^* generated by the columns of Δ such that $\mathcal{B}_1^*$ can be determined via elementary column operations on the matrix Δ.

2. Each u_i^* is a linear combination of the columns of Δ. For $1 \leq i \leq n$, let u_i denote the same linear combination of

the corresponding columns of $B(f)$. Then $\mathcal{B}_1 = \{u_1,\ldots,u_n\}$ is a basis for the D-module X_f.

3. $[u_1^*,\ldots,u_n^*]^{-1} \cdot \Gamma = {}_{S_1}[\beta]_{\mathcal{B}_1}$

4. ${}_{\mathcal{B}_1^*}[\sigma_p(\Phi) \cdot \Phi^{-1}]_{\mathcal{B}_1^*} = [u_1^*,\ldots,u_n^*]^{-1} \cdot \sigma_p(\Phi) \cdot \Phi^{-1}$

$\cdot\ [u_1^*,\ldots,u_n^*] = {}_{\mathcal{B}_1}[\sigma_p]_{\mathcal{B}_1}$

5. $\Lambda \cdot \Phi^{-1} \cdot [u_1^*,\ldots,u_n^*] = {}_{\mathcal{B}_1}[\gamma_p]_{S_2}$

Therefore, (H,F,G) is a matrix realization of f over D.

Proof. We first show that X_f^* is a free D-module of rank n. Being a finitely generated submodule of D^n, X_f^* is free since D is a Bézout domain. Moreover, since $\Phi^{(1)},\ldots,\Phi^{(n)}$ are linearly independent over D, the rank of X_f^* is at least n. But $\mathrm{rank}_L(B(f)) = n$ implies that the rank of X_f^* is at most n, as well.

Since $\det(\Phi) \neq 0$, the rows of Δ are linearly independent over L and *a fortiori* over D. The classical procedure over Bézout domains for selecting a basis for the column module X_f^* is given in steps (2)-(5) of the algorithm above. To see that $\mathcal{B}_1^* = \{u_1^*,\ldots,u_n^*\}$ is a basis for X_f^*, note that each u_j^* is zero in the first j - 1 coordinates, and that the jth coordinate of u_j^* must be nonzero; otherwise Δ_j, which is obtained from Δ by elementary column operations, would have rank less than n - j. To see that the vectors span the D-module X_f^* we have only to see that each column of Δ is a D-linear combination of u_1^*, $\ldots,u_n^*$. We illustrate the argument. Let u^* be a column of Δ. If b_1 is its first entry, then there exists an element $d_1 \in D$ so that $b_1 = d_1a_1$. Consider $u^* - d_1u_1^*$. Its first entry is 0 and its second entry is a multiple of a_2, say d_2a_2. Then $u^* - d_1u_1^* - d_2u_2^*$ has its first two entries 0, etc. Eventually, we arrive at $u^* - d_1u_1^* - \cdots - d_nu_n^* = 0$ and hence that $u^* = \sum_{i=1}^n d_iu_i^*$. This proves (1).

(2): By Theorem 4.17, since f is realizable over L, it must be realizable over D. Indeed, the proof of Theorem 4.17 shows that the recurrence relation for the family $\{A_i\}$ over L has its coefficients in D. Thus, X_f is generated over D by the columns of $B_{\infty,n}$. Moreover, since X_f is a finitely generated submodule of a free D-module, X_f is free. Now, $\text{rank}_L(B(f)) = n < \infty$ implies that X_f can be free of rank at most n, while $u_1^*,\ldots,u_n^*$ linearly independent implies that $u_1,\ldots,u_n$ are also linearly independent. Thus, $\mathcal{B}_1 = \{u_1,\ldots,u_n\}$ is a basis for the L-vector space spanned by the columns of B(f) and so each column of $B_{\infty,n}$ is a unique linear combination of $u_1,\ldots,u_n$. Since $\mathcal{B}_1^*$ is a basis for X_f^*, the coefficients of that linear combination must belong to D and so $u_1,\ldots,u_n$ also span X_f. This proves (2).

(3): First, notice that if $u \in X_f^*$ with $u = \sum_{i=1}^n d_iu_i^*$, then

$$[u_1^*,\ldots,u_n^*]^{-1}u = \sum_{i=1}^{n} d_i I^{(i)}$$

where $I^{(i)}$ stands for the ith column of the $n \times n$ identity matrix—that is, $[u_1^*,\ldots,u_n^*]^{-1}u$ gives the coordinate vector of u with respect to the basis $\mathcal{B}_1^*$. Therefore, $[u_1^*,\ldots,u_n^*]^{-1}\Gamma^{(j)}$ equals the coordinate vector of the jth column of Γ with respect to $\mathcal{B}_1^*$, which equals the coordinate vector of $\beta(\varepsilon_j)$ with respect to $\mathcal{B}_1$, as noted in the proof of (2). Hence,

$$[u_1^*,\ldots,u_n^*]^{-1} \cdot \Gamma = {}_{S_1}[\beta]_{\mathcal{B}_1}$$

(4): The first equality follows from the standard change of basis theorem. To see the second equality, let $[\beta_{ij}]$ be the matrix of σ_p with respect to $\mathcal{B}_1$; i.e.,

$$\sigma_p(u_j) = \sum_{i=1}^{n} \beta_{ij} u_i, \quad 1 \leq j \leq n$$

where the $\beta_{ij} \in D$. Let $\Phi^{(j)*}$ be the columns of the matrix Φ with $\Phi^{(j)}$ the corresponding long columns of $B(f)$. By the proof of Theorem 4.5, the $\Phi^{(j)}$, $1 \leq j \leq n$, form a basis over L for the column space of $B(f)$. Let

$$u_j = \sum_{i=1}^{n} \gamma_{ij} \Phi^{(i)}, \quad 1 \leq j \leq n$$

where the $\gamma_{ij} \in L$. Let $P : D^\infty \longrightarrow D^n$ be the transformation which projects an element of D^∞ to the rows which correspond to the rows of Φ. Then

$$\sigma_p(u_j) = \sum_{i=1}^{n} \gamma_{ij} \sigma_p(\Phi^{(i)})$$

and

$$\sum_{i=1}^{n} \beta_{ij} u_i^* = P\sigma_p(u_j) = \sum_{i=1}^{n} \gamma_{ij} \sigma_p(\Phi)^{(i)}$$

Hence

$$\begin{aligned}
\sigma_p(\Phi) \cdot \Phi^{-1} \cdot [u_1^*, \ldots, u_n^*] \cdot I^{(j)} &= \sigma_p(\Phi) \cdot \Phi^{-1} \cdot u_j^* \\
&= \sigma_p(\Phi) \cdot \left[\sum_{i=1}^{n} \gamma_{ij} \Phi^{-1} \cdot (\Phi^{(i)*}) \right] \\
&= \sum_{i=1}^{n} \gamma_{ij} \sigma_p(\Phi)^{(i)} \\
&= \sum_{i=1}^{n} \beta_{ij} u_i^*
\end{aligned}$$

It follows that $[\beta_{ij}]$ is the matrix

$$[u_1^*,\ldots,u_n^*]^{-1} \cdot \sigma_p(\Phi) \cdot \Phi^{-1} \cdot [u_1^*,\ldots,u_n^*]$$

(5): For $1 \leq j \leq n$, let

$$u_j^* = \sum_{i=1}^{n} \alpha_{ij}\Phi^{(i)*}$$

Then

$$\Lambda \cdot \Phi^{-1} \cdot [u_1^*,\ldots,u_n^*] \cdot e_j = \sum_{i=1}^{n} \alpha_{ij}\Lambda \cdot e_i$$

But $\Lambda \cdot e_i$ is merely the "chop" of $\Phi^{(i)}$ and

$$_{B_1}[\gamma_p]_{S_2} \cdot e_j = \gamma_p(u_j)$$

As we have seen,

$$u_j = \sum_{i=1}^{n} \alpha_{ij}\Phi^{(i)}$$

so that

$$\gamma_p(u_j) = \sum_{i=1}^{n} \alpha_{ij}\Lambda \cdot e_i$$

This completes the proof.

As we have stated before, all these results assume that we know that $\mathrm{rank}_L(B(f))$ is finite and hence that f is realizable.

4.3. RATIONALITY AND DESCENDING REALIZABILITY FROM THE QUOTIENT FIELD

Let D be an integral domain with quotient field L and let $f = \{A_i\}$ be an i/o map over D. If B(f) denotes the Hankel matrix of f, then we know that f is realizable over L if and only if $\mathrm{rank}_L(B(f))$ is finite. Since we wish to realize f over D, we would like to know conditions on D or conditions on the relationship between D and L which are sufficient to guarantee that realizability over L implies realizability over D. In this section we give such conditions. In fact, we are even able to characterize such integral domains and we do so in Theorem 4.17. We first prove a very nice special case.

THEOREM 4.11. Let D be an integral domain with quotient field L and let $f = \{A_i\}$ be an i/o map over D. Denote by X_f the D-module generated by the columns of the Hankel matrix B(f). If $\mathrm{rank}_L(B(f)) = n < \infty$, then

1. X_f is contained in a finitely generated D-module; in fact, X_f is contained in a D-module requiring only n generators.

2. If D is a noetherian domain, then f is realizable over D.

Proof. (1): Let Φ be an $n \times n$ submatrix of B(f) with $d = \det(\Phi) \neq 0$. As we've seen before, if $u_1^*,\ldots,u_n^*$ are the columns of Φ, then the long columns $u_1,\ldots,u_n$ of B(f) corresponding to $u_1^*,\ldots,u_n^*$ form a basis for the L-vector space generated by the columns of B(f). Moreover, given a column u of B(f), we can determine which linear combination of $u_1,\ldots,u_n$ it is merely by seeing which linear combination of $u_1^*,\ldots,u_n^*$ gives u^*. To do this, we seek elements $x_1,\ldots,x_n \in L$ so that $x_1u_1^* + \cdots + x_nu_n^* = u^*$. Now this is a system of n equations in n unknowns over L having Φ as coefficient matrix. By Cramer's Rule, we have that

$$x_i = \det([u_1^*,\ldots,u^*,\ldots,u_n^*])/d$$

for $1 \leq i \leq n$. Thus, for each u and each x_i the same d suffices. Therefore,

$$u = \sum_{i=1}^{n} (dx_i)\left(\frac{u_i}{d}\right)$$

and it follows that

$$X_f \subseteq D \cdot \left(\frac{u_1}{d}\right) + \cdots + D \cdot \left(\frac{u_n}{d}\right)$$

a D-module needing only n generators.

(2): If D is noetherian, then X_f, being a submodule of a finitely generated D-module, is itself finitely generated. The result follows from Theorem 4.4.

REMARK. Even though X_f is a submodule of a D-module with n generators, it does not follow that X_f is generated by n or fewer elements. Thus, we cannot conclude that f is n-realizable over D.

The key idea in the remainder of this section is that of associating to each i/o map f a formal power series. More precisely, if $f = \{A_i\}_1^{\infty}$ is a $p \times m$ i/o map over a commutative ring R, we form the $p \times m$ matrix

$$\left(\sum_{i=1}^{\infty} (A_i)_{jk} X^{-i}\right) \text{ for } 1 \leq j \leq p,\ 1 \leq k \leq m$$

a matrix with entries in the formal power series ring $R[[X^{-1}]]$. Although it may seem strange to use the indeterminate X^{-1} instead of X, it is, in fact, quite natural here for historical and mathematical reasons. Our first task is to relate this idea to the realizability of f. If

$$\alpha = \sum_{i=1}^{\infty} r_i X^{-i} \in X^{-1}R[[X^{-1}]]$$

call α *rational* if there exist polynomials P, Q $\in$ R[X], Q monic, such that α = P/Q. This equality is meaningful in the ring

$$R((X^{-1})) = \left\{ \sum_{i \geq n_0} a_i X^{-i} \middle| a_i \in R,\ n_0 \in Z \right\}$$

for if $X^n + \sum_{i=0}^{n-1} b_i X^i$ is a monic polynomial in R[X], then

$$X^n + \sum_{i=0}^{n-1} b_i X^i = X^n \cdot \left(1 + \sum_{i=0}^{n-1} b_i X^{i-n} \right)$$

Since $1 + \sum_{i=0}^{n-1} b_i X^{i-n}$ is a unit in $R[[X^{-1}]]$, $X^n \cdot (1 + \sum_{i=0}^{n-1} b_i X^{i-n})$ is not a zero divisor of $R((X^{-1}))$ and we can write $P/(X^n + \sum_{i=0}^{n-1} b_i X^i)$. If $f = \{A_i\}_1^\infty$ is a p $\times$ m i/o map over R, call f *rational* if each of the entries of the associated matrix of power series is rational.

Call a polynomial $p = \sum_{j=0}^n b_j X^j$ a *recurrence polynomial* for a sequence $\{a_i\}_1^\infty$ if $\sum_{j=0}^n a_{j+k} b_j = 0$ for all $k \geq 1$. If p is monic, this implies that $a_{n+k} = \sum_{j=0}^{n-1} (-b_j) a_{j+k}$ for $k \geq 1$, and hence, that $\{a_i\}_1^\infty$ is (n + 1)-recurrent.

We begin with the following result.

THEOREM 4.12. Let R be a commutative ring with X an indeterminate and let

$$R((X^{-1})) = \left\{ \sum_{i \geq n_0}^{\infty} a_i X^{-i} \middle| a_i \in R,\ n_0 \in Z \right\}$$

be the ring of Laurent series in X^{-1}. Let

$$\sum_{i=1}^{\infty} a_i X^{-i} \in X^{-1} \cdot R[[X^{-1}]]$$

and

$$\sum_{j=0}^{n} b_j X^j \in R[X]$$

Then

1. $$\left(\sum_{i=1}^{\infty} a_i X^{-i}\right) \cdot \left(\sum_{j=0}^{n} b_j X^j\right) = \sum_{j=1}^{n} \sum_{i=1}^{j} a_i b_j X^{j-i}$$

$$+ \sum_{k=1}^{\infty} \left(\sum_{j=0}^{n} a_{j+k} b_j\right) \cdot X^{-k}$$

where

$$\sum_{j=1}^{n} \sum_{i=1}^{j} a_i b_j X^{j-i}$$

is a polynomial in R[X] of degree at most $n - 1$ and

$$\sum_{k=1}^{\infty} \left(\sum_{j=0}^{n} a_{j+k} b_j\right) \cdot X^{-k}$$

is an element of $X^{-1} \cdot R[[X^{-1}]]$.

2. The following are equivalent.

a. $(\sum_{i=1}^{\infty} a_i X^{-i})(\sum_{j=1}^{n} b_j X^j) \in R[X]$ for some polynomial $\sum_{j=1}^{n} b_j X^j \in R[X]$.

b. $\sum_{j=0}^{n} b_j X^j$ is a recurrence polynomial for the sequence $\{a_i\}_1^{\infty}$. (Here, we do not require of a recurrence polynomial that it be monic.)

In the monic case, that is, if $b_n = 1$, then (a) and (b) are equivalent to

c. $\sum_{i=1}^{\infty} a_i X^{-i}$ is rational.

3. If $\{a_i\}_1^{\infty}$ is a recurrent sequence of elements of R, then the set of recurrence polynomials satisfied by $\{a_i\}$ is an ideal of R[X].

Proof. (1): We write out this calculation in detail.

$$\left(\sum_{i=1}^{\infty} a_i X^{-i}\right)\left(\sum_{j=0}^{n} b_j X^j\right) = \sum_{i=1}^{\infty}\left[a_i X^{-i}\left(\sum_{j=0}^{n} b_j X^j\right)\right]$$

$$= \sum_{i=1}^{\infty}\sum_{j=0}^{n} a_i b_j X^{j-i} = \sum_{j=0}^{n}\sum_{i=1}^{\infty} a_i b_j X^{j-i}$$

$$= \sum_{j=0}^{n}\left[\sum_{i=1}^{j} a_i b_j X^{j-i} + \sum_{i=j+1}^{\infty} a_i b_j X^{j-i}\right]$$

$$= \sum_{j=1}^{n}\sum_{i=1}^{j} a_i b_j X^{j-i} + \sum_{j=0}^{n}\sum_{i=j+1}^{\infty} a_i b_j X^{j-i}$$

which becomes, if we set $i = j + k$ in the second term,

$$= \sum_{j=1}^{n}\sum_{i=1}^{j} a_i b_j X^{j-i} + \sum_{k=1}^{\infty}\left(\sum_{j=0}^{n} a_{j+k} b_j\right) X^{-k}$$

This proves (1).

(2): The equivalence of (a) and (b) is clear from (1). Indeed, if $(\sum_{i=1}^{\infty} a_i X^{-i})(\sum_{j=0}^{n} b_j X^j) \in R[X]$, then

$$\left(\sum_{i=1}^{\infty} a_i X^{-i}\right)\left(\sum_{j=0}^{n} b_j X^j\right) = \sum_{j=1}^{n}\sum_{i=1}^{j} a_i b_j X^{j-i}$$

In the monic case, since $X^n + \sum_{j=0}^{n-1} b_j X^j$ is not a zero divisor in $R((X^{-1}))$, if (a) holds then

$$\left(\sum_{i=1}^{\infty} a_i X^{-i}\right) = \left(\sum_{j=0}^{n}\sum_{i=1}^{j} a_i b_j X^{j-i}\right) \Big/ \left(X^n + \sum_{j=0}^{n-1} b_j X^j\right)$$

that is, $\sum_{i=1}^{\infty} a_i X^{-i}$ is rational. On the other hand, if (c) holds with $\sum_{i=1}^{\infty} a_i X^{-i} = P/Q$, $P, Q \in R[X]$, Q monic, then $(\sum_{i=1}^{\infty} a_i X^{-i}) \cdot Q = P \in R[X]$ and (a) holds.

(3): Suppose that $\{a_i\}$ is recurrent and consider the power series $\sum_{i=1}^{\infty} a_i X^{-i}$. Now make repeated use of (2). If g and h are recurrence polynomials for $\{a_i\}$, then $(\sum_{i=1}^{\infty} a_i X^{-i}) \cdot$

$\in R[X]$ and $(\sum_{i=1}^{\infty} a_i X^{-i}) \cdot h \in R[X]$. Thus, $(\sum_{i=1}^{\infty} a_i X^{-i}) \cdot (g - h) \in R[X]$ and $g - h$ is a recurrence polynomial for $\{a_i\}$. Moreover, if g is a recurrence polynomial for $\{a_i\}$ and $h \in R[X]$ then $[(\sum_{i=1}^{\infty} a_i X^{-i}) \cdot g] \cdot h \in R[X]$ and $g \cdot h$ is a recurrence polynomial for $\{a_i\}$. This completes the proof.

We can now give the theorem relating realizability and rationality.

THEOREM 4.13. Let R be a commutative ring with $f = \{A_i\}$ a $p \times m$ i/o map over R. Then f is realizable if and only if f is rational. Moreover, suppose that f is rational and that the notation is

$$\left(\sum_{i=1}^{\infty} (A_i)_{jk} X^{-i} \right) = (P_{jk}/Q_{jk})$$

for polynomials P_{jk}, Q_{jk} in $R[X]$ with Q_{jk} monic. If n is the degree of a monic polynomial of smallest degree which is a multiple of each Q_{jk}, then f is $(n + 1)$-recurrent—that is, satisfies a monic recurrence polynomial of degree n.

Proof. In the notation of the theorem, f is rational if and only if

$$(P_{jk}/Q_{jk}) \cdot Q = \left(\sum_{i=1}^{\infty} (A_i)_{jk} X^{-i} \right) \cdot Q$$

is an element of $R[X]$ for $1 \leq j \leq p$, $1 \leq k \leq m$. This last equality holds if and only if Q is a monic recurrence polynomial for each of the sequences $\{(A_i)_{jk}\}_1^{\infty}$, which is equivalent to saying that Q is a monic recurrence polynomial for the i/o map f, which amounts to saying that f is realizable by Theorem 4.1.

We pause momentarily to refresh the reader on some notions we will need throughout the remainder of this section. Let D be an integral domain with quotient field L and let F be a field containing L. An element $\alpha \in F$ is said to be *almost integral* over D if the ring $D[\alpha]$ is contained in a finitely generated D-module. If $\alpha \in L$, this is equivalent to saying that there exists a nonzero element $d \in D$ such that $d\alpha^i \in D$ for $i \geq 0$. We say that D is *completely integrally closed* if each element of L that is almost integral over D is in D. An element $\beta \in F$ is said to be *integral* over D if the ring $D[\beta]$ is a finitely generated D-module or, equivalently, if β is a root of a monic polynomial with coefficients in D. We say that D is *integrally closed* if each element of L that is integral over D is in D. Clearly, any completely integrally closed domain is integrally closed. A unique factorization domain is completely integrally closed and therefore the rings Z and $L[X_1,\ldots,X_n]$ are completely integrally closed, where L is any field. A Bézout domain is integrally closed but need not be completely integrally closed. (See Chapter 1 for a more in-depth discussion of these ideas.) We return now to the main goal of this section.

We are trying to answer the question: For which integral domains D with quotient field L is it true that each i/o map over D which is realizable over L is realizable over D? Now by Theorem 4.13, an i/o map over D is realizable over D if and only if the associated power series in X^{-1} is rational over D. Since a matrix of power series is rational if and only if each entry is rational, a discussion of the scalar case suffices. Thus, if $f = \{a_i\}_1^\infty$ is a scalar i/o map over D, then $\alpha = \sum_{i=1}^\infty a_i X^{-i}$ is its associated power series. If f is rational over L, then $\alpha = P/Q$ where $P,Q \in L[X]$ and Q is monic. Defining

$$Q^* = Q/X^{\max\{\deg P,\ \deg Q\}}$$

and P^* analogously, we have that $P^*, Q^* \in L[X^{-1}]$ and that $\alpha = P^*/Q^*$ where the coefficient of the lowest degree term in Q^* is one. Now if Y is an indeterminate over D, let D(Y) denote the ring

$$\{g/h \mid g,h \in D[Y] \text{ and the coefficient of the lowest degree term in } h \text{ is one}\}$$

If we knew that

$$D[[X^{-1}]] \cap L(X^{-1}) = D(X^{-1})$$

then redividing by an appropriate power of X, we would have that $\alpha = g/h$, where $g,h \in D[X]$ and h is monic—that is, f is rational over D. Therefore, we seek to know when $D[[X]] \cap L(X) = D(X)$. The following theorem formalizes this and gives a purely ring-theoretic answer to the original question. It is the fundamental theorem of this section.

In the theorem we will use the standard notation for denoting a commutative ring localized at a multiplicatively closed set. Namely, if R is a commutative ring with S a multiplicatively closed subset of R, then $R_S = \{a/b \mid a \in R, b \in S\}$ denotes the ring of fractions of R with respect to the set S.

THEOREM 4.14. Let D be an integral domain with quotient field L. The following are equivalent.

1. Each i/o map over D that is realizable over L is realizable over D.

2. $(D[[X^{-1}]])_{\{X^{-i}\}_0^\infty} \cap L(X) = \{g/h \mid g,h \in D[X],\ h \text{ monic}\}$

3. $(D[[X^{-1}]])_{\{X^{-i}\}_0^\infty} \cap L(X^{-1}) = D(X^{-1})$

4. Each element of L that is almost integral over D is integral over D.

Proof. (1) ⟹ (2): We first note that

$$(D[[X^{-1}]])_{\{X^{-i}\}} = \left\{\sum_{i \geq n_0} a_i X^{-i} \mid n_0 \in Z,\ a_i \in D\right\}$$

Thus, it is clear that

$$\{g/h \mid g,h \in D[X],\ h \text{ monic}\} \subseteq (D[[X^{-1}]])_{\{X^{-i}\}} \cap L(X)$$

Conversely, if

$$\sum_{i \geq n_0} a_i X^{-i} \in (D[[X^{-1}]])_{\{X^{-i}\}} \cap L(X)$$

then

$$X^{n_0 - 1}\left(\sum_{i \geq n_0} a_i X^{-i}\right) \in (D[[X^{-1}]]) \cap L(X)$$

Therefore, the scalar i/o map $\{a_i\}_{n_0}^{\infty}$ is realizable over L. By hypothesis, $\{a_i\}$ is realizable over D and by Theorem 4.13,

$$X^{n_0 - 1} \cdot \left(\sum_{i \geq n_0} a_i X^{-i}\right) = \frac{g}{h}$$

for some monic polynomial $h \in D[X]$. Thus,

$$\sum_{i \geq n_0} a_i X^{-i} = X^{-n_0 + 1} \cdot \left(\frac{g}{h}\right)$$

as desired.

(2) ⟺ (3): This is obvious once one observes that $L(X) = L(X^{-1})$ and that $D(X^{-1})$ as defined above is equal to $\{g/h \mid g,h \in D[X],\ h \text{ monic}\}$.

(3) ⟹ (1): Let $f = \{A_i\}$ be an i/o map over D and let $\alpha = \sum_{i=1}^{\infty} a_i X^{-i}$ be an (arbitrary) one of the associated power series. If f is realizable over L, then $\sum_{i=1}^{\infty} a_i X^{-i} \in L(X)$ and so

$$\alpha \in (D[[X^{-1}]])_{\{X^{-i}\}} \cap L(X) = D(X^{-1})$$

which, as remarked above, is equal to $\{g/h \mid g,h \in D[X],\ h$ monic$\}$. Hence, α is rational over D and, since α was an arbitrary entry in the associated power series matrix for f, f is rational over D.

(3) <==> (4): Changing the variable from X^{-1} to X, the proof will be complete once we prove the following theorem.

THEOREM 4.15. Let D be an integral domain with quotient field L and let X be an indeterminate. Denote by D((X)) the ring $(D[[X]])_{\{X^i\}}$. Then $D((X)) \cap L(X) = D(X)$ if and only if each element of L almost integral over D is integral over D.

Proof. It is easy to see that each element $\alpha \in L(X)$ can be written uniquely in the form $\alpha = P_1/Q_1$ where P_1 and Q_1 are relatively prime polynomials in L[X] and where the coefficient of the lowest degree term of Q_1 is one. We call such a polynomial Q_1 *unitary* and we call the above representation of α the *irreducible unitary representation* of α. Note that if $\alpha = P/Q$, $P,Q \in L[X]$, Q unitary, then there exists a unitary polynomial $R \in L[X]$ such that $P = P_1 \cdot R$.

Before getting to the main body of the proof, we state and prove a useful lemma.

LEMMA 4.16. Let D be an integral domain with quotient field L. Given an element $\alpha \in L(X)$, the following are equivalent.

1. We can write $\alpha = P/Q$, where $P,Q \in L[X]$, Q is unitary and the coefficients of Q belong to D.

2. We can write $\alpha = P/Q$, where $P,Q \in L[X]$, Q is unitary and the coefficients of Q are integral over D.

3. The unique irreducible unitary representation P_1/Q_1 of α is such that the coefficients of Q_1 are integral over D.

Proof. (1) ==> (2): Obvious.

(2) ==> (3): By the above remark, $Q = Q_1 \cdot R$ where R is a unitary polynomial in L[X]. We need a little trick. Define a map

$$* : L[X] \longrightarrow L[X^{-1}] \text{ by } f^* = f/X^{\deg f}$$

If $f = g \cdot h$ for $g,h \in L[X]$, then

$$f^* = f/X^{\deg f} = gh/X^{\deg g + \deg h} = g^* \cdot h^*$$

Note also that the reverse map from $L[X^{-1}]$ to $L[X]$ does not alter the coefficients of a polynomial in $L[X^{-1}]$. Thus, * preserves factorizations and converts unitary polynomials in X to monic polynomials in X^{-1}. Therefore, $Q^* = Q_1^* \cdot R^*$ and each of Q^*, Q_1^*, and R^* is a monic polynomial in the indeterminate X^{-1}. A standard argument then shows that the coefficients of Q_1^*, and hence of Q_1, are integral over D. Since the same argument will be needed again, we sketch it here. If F is a splitting field of Q^* over L, then the roots of Q^* are integral over D since the coefficients of Q^* are integral over D. In particular, the roots of Q_1^* are integral over D. But the coefficients of Q_1^* are elementary symmetric functions in the roots of Q_1^* and so are integral over D.

It being obvious that (3) implies (2), we prove only that

(2) ⟹ (1): We first prove the following: If f is a monic polynomial in L[X] with coefficients integral over D, then there exists a monic polynomial $g \in L[X]$ such that $f \cdot g \in D[X]$. If F is a splitting field for f over L, then in F[X],

$$f = \prod_{i=1}^{n} (X - \alpha_i)$$

For $1 \leqslant i \leqslant n$, let h_i be a monic polynomial in D[X] having α_i as a root. Then the polynomial $g = (\prod h_i)/f$ fulfills the requirement. Indeed, since $\prod h_i$ and f belong to L[X] and f divides $\prod h_i$ in F[X], it follows easily that f divides $\prod h_i$ in L[X]. Thus, $g \in L[X]$.

In our case, we have that $P_1/Q_1 \in L(X)$ and is such that the coefficients of Q_1 are integral over D. Hence, the coefficients of Q_1^* are integral over D and Q_1^* is monic in $L[X^{-1}]$. By what we just proved, there is a monic polynomial R^* in $L[X^{-1}]$ such that $Q_1^* \cdot R^*$ has coefficients in D. Then $(P_1/Q_1) \cdot (R/R) = P_1/Q_1$ is such that the coefficients of $Q_1 \cdot R$ belong to D. This completes the proof of the lemma.

We return at last to the proof of Theorem 4.15. Since a power series in D[[X]] is invertible if and only if its constant term is invertible in D, we see that $D(X) \subseteq L(X) \cap D((X))$ Indeed, from Lemma 4.16 it follows that D(X) is the subset of $L(X) \cap D((X))$ consisting of those elements whose unique irreducible unitary representations have coefficients integral over D.

We next prove that the unique irreducible unitary representation of each element of $L(X) \cap D((X))$ has its coefficients almost integral over D. Thus, if each element of L almost integral over D is integral over D, it will follow that $L(X) \cap D((X)) = D(X)$. This will prove one half of the theorem.

Thus, let P_1/Q_1 be the unique irreducible unitary representation of an element of $L(X) \cap D((X))$. Since we are only interested in the coefficients of Q_1 we can, by multiplying by an appropriate power of X, assume that $P_1/Q_1 = P/Q \in D[[X]] \cap L(X)$ and that the constant term of Q is one. Firstly, consider the case when $Q = 1 - \lambda X$, $\lambda \in L$. We have that

$$\frac{P}{1 - \lambda X} = \sum_{j=0}^{\infty} a_j X^j, \quad a_j \in D$$

If the power series is actually a polynomial, then $P \in (1 - \lambda X)L[X]$, contrary to the fact that P_1 and Q_1 are relatively prime. Therefore, $P = (1 - \lambda X) \sum_{j=0}^{\infty} a_j X^j$. If $n > \deg P$, then it follows that $a_n - \lambda a_{n-1} = 0$. If we choose $m > \deg P$ such that $a_m \neq 0$, then $a_m \lambda = a_{m+1}$. By induction, $a_m \lambda^t = a_{m+t} \in D$ and so λ is almost integral over D.

Now consider the general case. The polynomial Q has its coefficients almost integral over D if and only if Q^* does. But Q^* is monic so that in a splitting field we can factor $Q^* = (X^{-1} - \lambda_1)\cdots(X^{-1} - \lambda_n)$. Since the coefficients of Q^* are polynomials in $\lambda_1,\ldots,\lambda_n$, it suffices to see that each λ_i is almost integral over D. Moreover, since $\lambda_1,\ldots,\lambda_n$ are algebraic over L, the quotient field of D, we can find by a standard argument a nonzero element $d \in D$ such that $d\lambda_1,\ldots,d\lambda_n$ are integral over D. Now pass to the integral domain $B = D[d\lambda_1,\ldots,d\lambda_n]$. Since B is a finitely generated D-module, if λ_i is almost integral over B, it is almost integral over D. Now $Q = (1 - \lambda_1 X)\cdots(1 - \lambda_n X)$ and $P/Q = \sum_{j=0}^{\infty} a_j X^j \in D[[X]]$. Multiply through by $d^{n-1}(1 - \lambda_2 X)\cdots(1 - \lambda_n X)$ to get $P/(1 - \lambda_1 X) \in B[[X]]$. Since $d\lambda_1 \in B$, it follows that λ_1 belongs to the quotient field of B and so we may apply the previous case to conclude that λ_1 is almost integral over B. Indeed, this argument shows that each λ_i is almost integral over B and hence over D.

Finally, suppose that $L(X) \cap D((X)) = D(X)$ and that t is an element of L which is almost integral over D. Let d be a nonzero element of D such that $dt^n \in D$ for all positive integers n and consider the fraction $\alpha = d/(1 - tX)$. Since $dt^n \in D$, it is easy to see that $d \cdot (1 - tX)^{-1} \in D[[X]]$ and hence that $\alpha \in L(X) \cap D((X)) = D(X)$. It follows from Lemma 4.16 that t is integral over D. This completes the proof of both Theorem 4.15 and Theorem 4.14.

Any completely integrally closed domain has the property given in Theorem 4.15, as does any integral domain whose integral closure is completely integrally closed. In fact, it is obvious that an integrally closed domain has the property if and only if it is completely integrally closed. In particular, a Bézout domain has the property if and only if it is completely integrally closed. As there exist Bézout domains that are not completely integrally closed [22, p. 194], there exist Bézout domains having i/o maps that are realizable over the quotient field but not over the domain itself.

Theorem 4.14 has an interesting corollary that was used in the proof of Theorem 4.10. Among other things, the corollary, Theorem 4.17, says that over a completely integrally closed domain D, the i/o maps realizable over the quotient field are realizable over D and in the single-input case the order of realizability over the domain is the same as over the quotient field.

THEOREM 4.17. Let D be a completely integrally closed domain having quotient field L and let f be an i/o map over D. If $\text{rank}_L(B(f)) = n < \infty$, then f is n-realizable over D. Moreover, the recurrence polynomial of f over L has its coefficients in D. In particular, f is (n + 1)-recurrent over D, and consequently, f is realizable over D of order at most mn.

Proof. Since $\text{rank}_L(B(f))$ is finite, f is realizable over L and, by Theorem 4.14, also realizable over D. Moreover, f satisfies a monic recurrence relation g with coefficients in D given, for example, by the characteristic polynomial of one of its realizations. Viewing f as a sequence over L, the set of recurrence polynomials for f is an ideal of L[X] and therefore is principal, generated by a monic polynomial $h \in L[X]$. Thus, $g = h \cdot k$ for some monic polynomial $k \in L[X]$. The standard argument we gave in the proof of Lemma 4.16 shows that the coefficients of h are integral over D and therefore belong to D.

4.4. CANONICAL SYSTEMS AND CANONICAL REALIZATIONS

In previous sections of this chapter we were concerned with the problem of finding a system to realize a given i/o map. In this section we shall concern ourselves with determining the best possible or most natural realization. In so doing, we shall tie together realizability, reachability and observability, and we shall also discuss what it means for two realizations to be isomorphic.

A system (H,F,G) over a commutative ring R is called *canonical* if it is both reachable and observable. Likewise, if f is an i/o map over R, then a realization of f is said to be *canonical* if it is both reachable and observable. Now this latter definition needs some amplification if the state module is not a free module, for we only treated reachability and observability for free system—that is, for systems with free state spaces. Consider the system (X,H,F,G) given diagramatically by

$$R^m \xrightarrow{G} X \xrightarrow{F} X \xrightarrow{H} R^p$$

The reachability map of the system is the R-homomorphism

$$\phi : \bigoplus_{i=0}^{\infty} R^m \longrightarrow X$$

given by $G, F \circ G, F^2 \circ G, \ldots$. The observability map of the system is the R-homomorphism $\tau : X \longrightarrow \prod_{i=0}^{\infty} R^p$ given by $H, H \circ F$, $H \circ F^2, \ldots$. Then the system is said to be *reachable* if ϕ is surjective and *observable* if τ is injective.

THEOREM 4.18. Let $f = \{A_i\}$ be an i/o map over a commutative ring R and let X_f denote the R-module generated by the columns of the Hankel matrix B(f). Then f is realizable if and only if X_f is finitely generated. Moreover, the "blow-up, shift, and chop" realization having X_f as state module is canonical.

Proof. The first assertion is merely a restatement of Theorem 4.4 and so we have only to prove the moreover claim.

The reachability map

$$\phi : \bigoplus_{i=0}^{\infty} R^m \longrightarrow X_f$$

is given by $\beta \oplus \sigma \circ \beta \oplus \sigma^2 \circ \beta \oplus \cdots$. Suppose that C is an elementary column from B(f), say C is the ith elementary column in the block column headed by A_n. Then $\phi(0,0,\ldots,\varepsilon_i,0,\ldots) = C$ where ε_i is in the (n - 1)st coordinate. Thus, each generator of X_f is in the image of ϕ and it follows that ϕ is surjective.

The observability map

$$\tau : X_f \longrightarrow \prod_{i=0}^{\infty} R^p$$

is given by γ, $\gamma \circ \sigma$, $\gamma \circ \sigma^2, \ldots$. Now τ is the restriction to X_f of the R-homomorphism

$$\tau^* : R^\infty \longrightarrow \prod_{i=0}^{\infty} R^p$$

defined in the obvious way. If $\tau^*(x) = 0$, then $x = 0$. Thus, since τ^* is injective on R^∞, its restriction τ is injective on X_f.

What sort of uniqueness is there for realizations of a given i/o map? In general, there is not very much, but we can say a little for canonical realizations. The reader should note that the proof of the following theorem is rather formal. Indeed, a similar remark holds for much of the remainder of this chapter.

THEOREM 4.19. Let R be a commutative ring with $f = \{A_i\}_1^\infty$ an i/o map over R. If (X,H,F,G) and (X_1,H_1,F_1,G_1) are two canonical realizations of f, then there is an isomorphism $\psi : X \longrightarrow X_1$ such that $\psi \circ G = G_1$, $H_1 \circ \psi = H$, and $\psi \circ F = F_1 \circ \psi$.

Proof. Assume the notation is the usual; that is, A_i is a $p \times m$ matrix for $i \geq 1$. Let Φ, Φ_1, and τ, τ_1 be the reachability and observability maps of (X,H,F,G) and (X_1,H_1,F_1,G_1), respectively. Then $\tau \circ \Phi$ and $\tau \circ \Phi_1$ are each homomorphisms from $\oplus_{i=0}^{\infty} R^m$ to $\prod_{i=0}^{\infty} R^p$. We claim that $\tau \circ \Phi = \tau_1 \circ \Phi_1$ and, indeed, that this does not depend upon the fact that the realizations are canonical. Inelegantly, if

$$u = (u_0,u_1,\ldots,u_t,0,0,\ldots) \in \bigoplus_{i=0}^{\infty} R^m$$

then

$$\begin{aligned}
(\tau \circ \Phi)(u) &= \tau(Gu_0 + FGu_1 + \cdots + F^tGu_t) \\
&= (HGu_0 + HFGu_1 + \cdots + HF^tGu_t, HFGu_0 + HF^2Gu_1 \\
&\qquad + \cdots + HF^{t+1}Gu_t, \ldots) \\
&= (A_1u_0 + A_2u_1 + \cdots + A_{t+1}u_t, A_2u_0 + A_3u_1 + \cdots \\
&\qquad + A_{t+2}u_t, \ldots) \\
&= (H_1G_1u_0 + H_1F_1G_1u_1 + \cdots + H_1F_1{}^tG_1u_t, H_1F_1G_1u_0 \\
&\qquad + H_1F_1^2G_1u_1 + \cdots + H_1F_1{}^{t+1}G_1u_t, \ldots) \\
&= \tau_1(G_1u_0 + F_1G_1u_1 + \cdots + F_1^tG_1u_t) \\
&= (\tau_1 \circ \Phi_1)(u)
\end{aligned}$$

and the claim is verified.

Thus, we are led to the following commutative diagram of R-modules

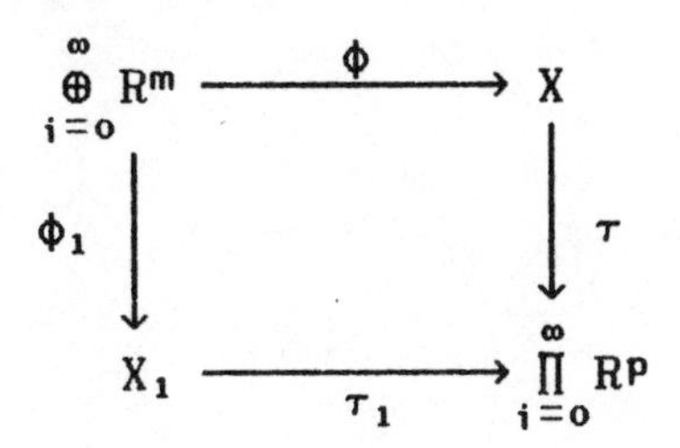

where ϕ and ϕ_1 are surjections and τ and τ_1 are injections. By means of some elementary diagram chasing, we can complete the proof. We will need the following lemma.

LEMMA. Let

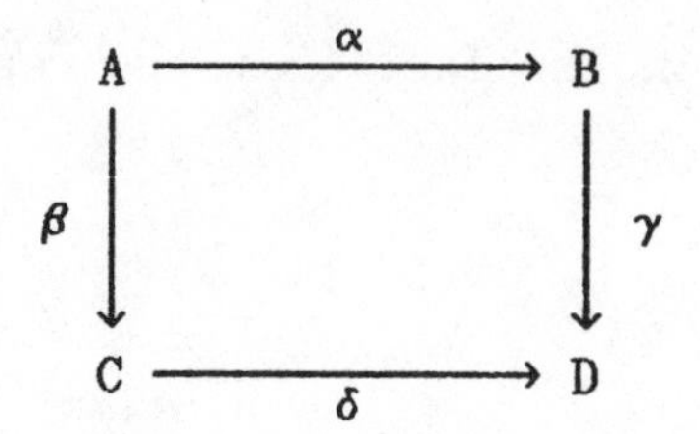

be a commutative diagram of R-modules with α surjective and δ injective. Then there exists an R-homomorphism $\rho : B \longrightarrow C$ such that $\rho \circ \alpha = \beta$ and $\gamma = \delta \circ \rho$.

Proof of the lemma. If $b \in B$, then there exists an element $a \in A$ such that $\alpha(a) = b$. Set $\rho(b) = \beta(a)$. To see that ρ is well defined, if $a' \in A$ is such that $\alpha(a') = b$, then

$$\gamma(b) = \gamma(\alpha(a)) = \gamma(\alpha(a')) = \delta(\beta(a)) = \delta(\beta(a'))$$

Since δ is injective, $\beta(a) = \beta(a') = \rho(b)$ as required.

Moreover, if $a \in A$, then $\beta(a) = \rho(\alpha(a))$ by the very definition of ρ. If $b \in B$ and if $a' \in A$ is such that $\alpha(a') = b$, then

$$(\delta \circ \rho)(b) = \delta(\beta(a')) = \gamma(\alpha(a')) = \gamma(b)$$

and the lemma has been established.

Returning to the proof of the theorem we may, by the lemma, fill in our diagram with an R-homomorphism ψ as indicated below.

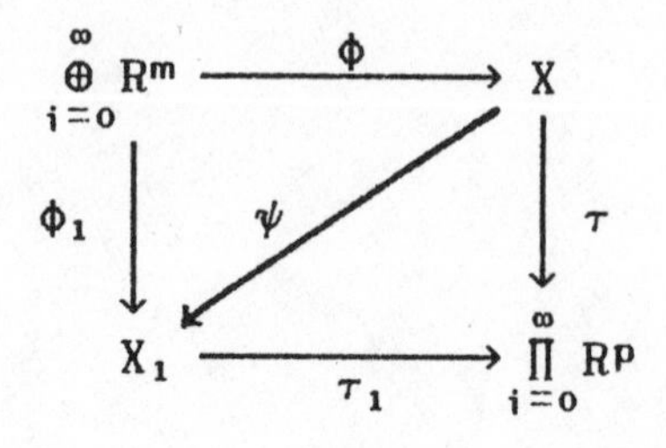

Here, $\Phi_1 = \psi \circ \Phi$ and $\tau_1 \circ \psi = \tau$. Since Φ_1 is surjective, so is ψ, and since τ is injective, so is ψ. Therefore, ψ is an isomorphism.

Since $\Phi_1 = \psi \circ \Phi$, we have that

$$G_1x = \Phi_1(x,0,0,\ldots) = (\psi \circ \Phi)(x,0,0,\ldots) = \psi(Gx)$$

for all x in R^m. Hence, $\psi \circ G = G_1$. Similarly, the equation $\tau_1 \circ \psi = \tau$ implies that $H_1 \circ \psi = H$. For any two realizations of f, it follows from the definitions that $\tau \circ F \circ \Phi = \tau_1 \circ F_1 \circ \Phi_1$. Thus, in our case

$$(\tau_1 \circ \psi) \circ F \circ \Phi = \tau_1 \circ F_1 \circ (\psi \circ \Phi)$$

Since τ_1 is injective and Φ is surjective, it follows that $\Phi \circ F = F_1 \circ \psi$. This completes the proof of Theorem 4.19.

With Theorem 4.19 in mind, we define the notion of a morphism between a pair of systems. Let (X,H,F,G) and (X_1,H_1,F_1,G_1) be two systems over a commutative ring R each having R^m as input module and R^p as output module. By a ***morphism*** from (X,H,F,G) to (X_1,H_1,F_1,G_1) we mean an R-homomorphism $\psi : X \to X_1$ making the following diagram commute.

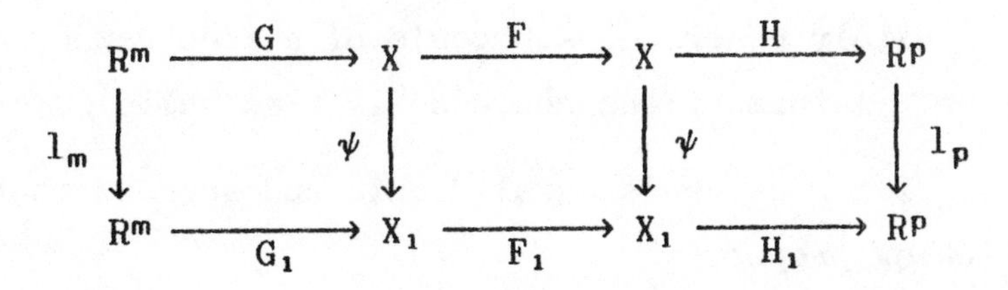

(Here, 1_m denotes the identity map on R^m and 1_p the identity on R^p.) If ψ is a bijection, we call ψ an isomorphism and we say that (X,H,F,G) and (X_1,H_1,F_1,G_1) are isomorphic.

With this terminology, Theorem 4.19 can be rephrased.

THEOREM 4.20. Let R be a commutative ring with $f = \{A_i\}_i^\infty$ an i/o map over R. If (X,H,F,G) and (X_1,H_1,F_1,G_1) are two canonical realizations of f, then (X,H,F,G) and (X_1,H_1,F_1,G_1) are isomorphic.

We summarize in the following theorem much of the work of this chapter. The result is the natural ring-theoretic version of the classical theorem for fields.

THEOREM 4.21. Let D be a completely integrally closed Bézout domain with quotient field L and let f be an i/o map over D. Then f has a realization over D with finitely generated free state module if and only if $\text{rank}_L(B(f)) = n < \infty$, where B(f) denotes the Hankel matrix of f. Moreover, when $\text{rank}_L(B(f)) = n < \infty$, the state module of the canonical realization is D^n and the algorithm given in Theorem 4.10 yields the canonical realization.

As we have seen, given an i/o map f over a ring R, the column module X_f of the Hankel matrix of f is the state module of "the" canonical realization of f in the event that X_f is finitely generated. Thus, it can happen that the state module of a canonical realization of a realizable i/o map over R is not a free module. Of course, if D is a Bézout domain, then X_f, being a finitely generated submodule of a free D-module, is itself free. A form of the converse is also valid.

THEOREM 4.22. Let D be an integral domain and suppose that D has the following property:

> The state module of the canonical realization of every realizable i/o map over D is a free D-module.

Then D is a Bézout domain.

Proof. By virtue of (Chapter 1, Theorem 1.21), it suffices to show that each finitely generated torsion-free module over D is free. Thus, let E be a finitely generated torsion-free D-module and suppose that E is generated by m elements. We can map the free module D^m onto E in a natural way, leading to exact sequence

$$D^m \xrightarrow{\sigma} E \longrightarrow 0$$

Moreover, since D is an integral domain and since E is a finitely generated torsion-free D-module, we can embed E in a finitely generated free D-module [51, Lemma 4.31]. This leads to the exact sequence

$$0 \longrightarrow E \xrightarrow{\tau} D^P$$

Combining these, we have the following sequence (which is most definitely not exact):

$$D^m \xrightarrow{\sigma} E \xrightarrow{1_E} E \xrightarrow{\tau} D^P$$

Evidently, the system $(E, \tau, 1_E, \sigma)$ is canonical and has E as state module. In particular, E is the state module of the canonical realization of the i/o map $\{\tau \circ \sigma\}$ and, by hypothesis, E must be a free D-module. Hence D must be a Bézout domain.

Finally, in the general case, if we are willing to forego "canonicity," we can show that if an i/o map f is (finitely) realizable, then it has a realization with a free state module. The following theorem is a partial reformulation of Theorem 4.4 incorporating the free state module assertion.

THEOREM 4.23. Let $f = \{A_i\}$ be an i/o map over a commutative ring R. There exists a realization of f with a finitely generated state module if and only if X_f, the R-module generated by the columns of the Hankel matrix of f, is finitely generated. Moreover, if such a realization exists, then there exists a realization of f having a finitely generated free state module.

Proof. In light of Theorem 4.4, we have only to prove the final claim. Thus, f is realizable, say by (X,H,F,G) with X a finitely generated R-module. If X can be generated by n elements, then we can map R^n onto X, say by ψ. This leads us to the following diagram whose bottom row is exact.

$$\begin{array}{ccccc} & & R^n & & \\ & & \downarrow \psi & & \\ & & X & & \\ & & \downarrow F & & \\ R^n & \xrightarrow[\psi]{} & X & \rightarrow & 0 \end{array}$$

Since R^n is a projective R-module and since the bottom row is exact, there exists an R-homomorphism $F_1 : R^n \rightarrow R^n$ making the resulting diagram commute. Re-drawing the diagram we have

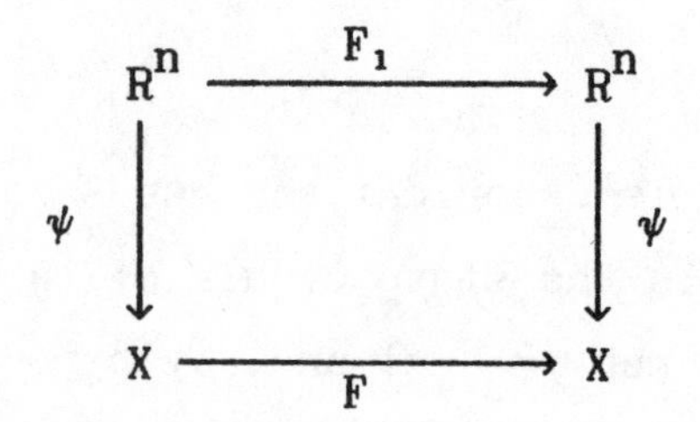

Suppose that R^m is the input space and R^p the output space for (X,H,F,G). We can enlarge the above picture as follows. Since R^m is projective, the diagram

$$\begin{array}{ccccc} & & R^m & & \\ & & \downarrow G & & \\ R^n & \xrightarrow[\psi]{} & X & \rightarrow & 0 \end{array}$$

can be completed by means of a homomorphism $G_1 : R^m \rightarrow R^n$ making the resulting diagram commute—that is, $\psi \circ G_1 = G$. Now, set $H_1 = H \circ \psi$. Combining commutative diagrams we are left with the following one.

$$\begin{array}{ccccccc} R^m & \xrightarrow{G_1} & R^n & \xrightarrow{F_1} & R^n & \xrightarrow{H_1} & R^p \\ 1_m \big\downarrow & & \psi \big\downarrow & & \big\downarrow \psi & & \big\downarrow 1_p \\ R^m & \xrightarrow[G]{} & X & \xrightarrow[F]{} & X & \xrightarrow[H]{} & R^p \end{array}$$

This says precisely that ψ is a morphism from (R^n, H_1, F_1, G_1) to (X,H,F,G). By the following lemma, since (X,H,F,G) realizes f, so does (R^n, H_1, F_1, G_1).

LEMMA. Let ψ be a morphism from the system (X,H,F,G) to the system (X_1, H_1, F_1, G_1). Then $H \circ F^i \circ G = H_1 \circ F_1^i \circ G_1$ for $i \geqslant 0$.

Proof. Since ψ is a morphism we have, suppressing the composition notation, that $\psi G = G_1$, $H_1\psi = H$, and $F_1\psi = \psi F$. Suppose by way of induction that $F_1^k\psi = \psi F^k$. Then $F_1^{k+1}\psi = F_1\psi F^k = \psi F^{k+1}$. So, we find that $H_1G_1 = H_1\psi G = HG$, and that for $k \geqslant 1$, $H_1F_1^k\psi = H_1\psi F^k = HF^k$ from which it follows that $H_1F_1^k\psi G = H_1F_1^kG_1 = HF^kG$.

Theorem 4.23 settles once and for all the issue of psuedo-realization introduced earlier in this chapter. That is, if an i/o map has a pseudo-realization with a finitely generated state module, then it has an actual realization with a finitely generated free state module. In particular, the maps H, F, and G are given by matrices.

EXERCISES

1. Consider the systems (C,A,B) and (H,F,G) over the reals, where

$$C = [0,0,1], \quad H = [0,1]$$

$$A = \begin{bmatrix} 0 & 0 & 0 \\ 1 & 0 & 0 \\ 0 & 1 & 0 \end{bmatrix}, \quad F = \begin{bmatrix} 0 & 0 \\ 1 & 0 \end{bmatrix}$$

$$B = \begin{bmatrix} 0 \\ 1 \\ 0 \end{bmatrix}, \quad G = \begin{bmatrix} 1 \\ 0 \end{bmatrix}$$

Prove that the two systems are observable realizations of the same i/o map.

2. Let (C,A,B) and (C_1,A_1,B_1) be two systems over a ring R where

$$C = [C_1 \quad 0]$$

$$A = \begin{bmatrix} A_1 & 0 \\ A_2 & A_3 \end{bmatrix}$$

$$B = \begin{bmatrix} B_1 \\ B_2 \end{bmatrix}$$

Prove that the two systems are realizations of the same i/o map.

3. (Finite Hankel Matrices) Suppose that L is a field and $a_i \in L$ for $i = 0,\dots,2N - 2$. Let

$$B_N(f) = \begin{bmatrix} a_0 & a_1 & \cdots & a_{N-1} \\ a_1 & a_2 & & a_N \\ \vdots & \vdots & & \vdots \\ a_{N-1} & a_{N-2} & \cdots & a_{2N-2} \end{bmatrix}$$

Let $D_1, D_2, \dots, D_N$ denote the successive principal minors of $B_N(f)$. Show that if the first k rows of $B_N(f)$ are linearly independent and if the first k + 1 rows of $B_N(f)$ are linearly dependent, then $D_k \neq 0$.

4. In Theorem 4.6, for the scalar case $m = p = 1$, it was shown that, for an infinite Hankel matrix $B(f)$, $\text{rank}(B(f)) = n$ implies that $D_n \neq 0$. Give an example to show that this is not necessarily true for a finite Hankel matrix.

5. Let $B(f)$ be the infinite Hankel matrix corresponding to a scalar i/o map $f = \{a_i\}_{i=0}^{\infty}$ over a field L. Prove that $B(f)$

has rank n if and only if there exist elements $r_1,\dots,r_n \in L$ such that

$$a_k = r_1 a_{k-1} + r_2 a_{k-2} + \cdots + r_n a_{k-n} \quad \text{for} \quad k = n, n+1, \dots \qquad (*)$$

and n is the smallest integer for which (*) is true.

6. Let f and B(f) be as in Exercise 5. Show directly that B(f) has finite rank if and only if the f is rational.

7. Let $f = \{A_i\}_{i=1}^{\infty}$ be an i/o map over an arbitrary integral domain and suppose f is (n + 1)-recurrent. Prove that the first n block columns of B(f) contain a basis $c_1, c_2, \dots, c_k$ for X_f over the quotient field. If n = 2, prove that the short columns $\hat{c}_1, \hat{c}_2, \dots, \hat{c}_k$ formed by the parts of $c_1, c_2, \dots, c_k$ respectively lying in the first 2 block rows are also linearly independent. (This is surely true for n > 2, but the notation becomes difficult.)

8. Consider the transfer function in Exercise 5 of the Introduction. Find a realization of the corresponding i/o map and show that the corresponding system of delay differential equations is

$$\frac{dx_1}{dt} = -x_1(t-1) + x_2(t) + x_2(t-1) + x_2(t-2) + u_1(t) + u_2(t-1)$$

$$\frac{dx_2}{dt} = x_2(t) + u_2(t)$$

$$y_1(t) = x_1(t)$$

$$y_2(t) = -x_1(t) + x_2(t-1)$$

9. Prove directly that the realization in the previous exercise is canonical.

10. Let R = k[X,Y,Z] be the polynomial ring in three variables over a field k. Let $f = \{A_i\}_{i=1}^{\infty}$ be the i/o map given by

$$A_1 = \begin{bmatrix} X & Z & 0 \\ Y & 0 & -Z \\ 0 & -Y & -X \end{bmatrix}$$

and $A_j = 0$, $j \geq 2$. Use Theorem 4.9 to find a realization of order two of f over the quotient field. (In [53] it is shown that f does not have a realization over R of order two.)

11. Let (H,F,G) be a system defined over a field k. Let L be the image in k^n of the reachability matrix of the system. Let L′ be an algebraic complement of L in k^n. Show that, with respect to the decomposition $k^n = L \oplus L'$, the system decomposes as

$$H = [H_1, H_2]$$

$$F = \begin{bmatrix} F_1 & F_2 \\ 0 & F_3 \end{bmatrix}$$

$$G = \begin{bmatrix} G_1 \\ 0 \end{bmatrix}$$

Show that (H,F,G) and (H_1,F_1,G_1) realize the same i/o map.

12. Let (H,F,G) be a system defined over a field k. Let

$$L = \{x \in k^n \mid HF^jx = 0 \quad \text{for all } j\}$$

Let L′ be an algebraic complement of L in k^n. Show that, with respect to the decomposition $k^n = L \oplus L'$, the system decomposes as

$$H = [0, H_1]$$

$$F = \begin{bmatrix} F_1 & F_2 \\ 0 & F_3 \end{bmatrix}$$

$$G = \begin{bmatrix} G_1 \\ G_2 \end{bmatrix}$$

Show that (H,F,G) and (H_1,F_3,G_2) realize the same i/o map.

13. Let f be an i/o map over a field k. Prove that any two realizations of f of minimum order are isomorphic.

NOTES AND REMARKS

We learned Theorem 4.1 from Sontag's survey article [59, Section 3]. For the case of R a field, it appears in [31, Lemmas 11.5 and 11.7], where it is attributed to Ho [29] and, independently, to Youle and Tissi [67]. We learned Theorem 4.2 from [53, p. 68] and [59, p. 18], where it is attributed to the dissertations of Rouchaleau [52] and Fliess [20]. Theorem 4.3 is from [18, p. 433]. We proved Theorem 4.4 as in [59, p. 18]. Theorem 4.5 is folklore.

Theorem 4.6 is stated in [59, p. 18] and a proof, in the single input case, is recorded in [49, Lemma 2]. Theorem 4.9 is from [57]. We learned Theorem 4.10 in the principal ideal domain case from [33, p. 20], where it is attributed to [52]. Kamen noted in [34, p. 874] that a realizable map over a Bézout domain could be minimally realized. We included completely integrally closed in the hypotheses of Theorem 4.10 so we could use Theorem 4.17 to conclude that the input/output map was realizable over D if it was realizable over the quotient field.

Theorem 4.11 is due to Rouchaleau, Wyman, and Kalman [54]. We learned it from [45]. The proof we give is from [59, Appendix I]. We learned 4.13 from [33] and [59]; it seems to go back to Kronecker, see [48, p. 424]. Our proof of Theorem 4.13 is a corollary of Theorem 4.12, which we formulated after studying [13]. Theorems 4.14 and 4.15, as well as Lemma 4.16, are from [13]. The "moreover" part of Theorem 4.17 is from [53, Lemma 3.2].

We learned Theorem 4.19 from [45, Theorem 4.4], where it is attributed to [1, Lemma 4.6]. Theorem 4.21 for fields is in [31]. Theorem 4.22 in the noetherian case is due to [53, Proposition 4.1]. Theorem 4.23 is from [18, Proposition 3.1].

Exercises 3, 4, 5, and 6 are from [21]. Exercise 8 (and Exercise 5 of the Introduction) are from [33]. Exercise 10 is from [53].

References

1. M. Arbib and E. Manes, Foundations of systems theory: decomposable systems, Automatica 10 (1974), 285-302.
2. M. Atiyah and I. Macdonald, Introduction to Commutative Algebra, Addison-Wesley, Reading, Massachusetts, 1969.
3. A. Balakrishnan, Elements of State Space Theory of Systems, Optimization Software, New York, 1983.
4. S. Barnett, Polynomial and Linear Control Systems, Marcel Dekker, New York, 1983.
5. R. Bellman and K. Cooke, Differential-Difference Equations, Academic Press, New York, 1963.
6. J. Brewer and D. Costa, Projective modules over some non-noetherian polynomial rings, Jour. Pure Appl. Algebra 13 (1978), 157-163.
7. J. Brewer, W. Heinzer and D. Lantz, The pole assignability property in polynomial rings over GCD-domains, preprint.
8. J. Brewer, D. Katz, and W. Ullery, On pole assignability over commutative rings, preprint.
9. J. Brewer, C. Naudé and G. Naudé, On Bézout domains, elementary divisor rings, and pole assignability, Comm. Algebra, to appear.
10. R. Bumby, E. Sontag, H. Sussmann and W. Vasconcelos, Remarks on the pole-shifting problem over rings, J. Pure Appl. Algebra 20 (1981), 113-127.
11. C. Byrnes, On the stabilizability of linear control systems depending on parameters, Proc. 18th IEEE Conf. Dec. and Control (1979), 233-236.

12. C. Byrnes, Realization theory and quadratic optimal controllers for systems defined over Banach and Frechet algebras, Proc. IEEE Conf. Dec. and Control (1980), 247-255.

13. P. Cahen and J. Chabert, Elements quasi-entiers et extensions de Fatou, Journal of Algebra 36 (1975), 185-192.

14. W. Ching and B. Wyman, Duality and the regulator problem for linear systems over commutative rings, J. Comput. System Sci. 14 (1977), 360-368.

15. J. Dieudonne, Foundations of Modern Analysis, Academic Press, New York, 1960.

16. R. Driver, Ordinary and Delay Differential Equations, Springer-Verlag, New York, 1977.

17. D. Dubois, A nullstellensatz for ordered fields, Ark. Mat. 8 (1969), 111-114.

18. S. Eilenberg, Automata, Languages and Machines, Vol. A, Academic Press, New York, 1974.

19. R. Eising, Pole assignment for systems over rings, Systems and Control Letters 2 (1982), 225-229.

20. M. Fliess, Sur certaines familles de series formelles, These de Doctorat d'Etat, University Paris VII, 1972.

21. F. Gantmacher, The Theory of Matrices, Vol. I and II, Chelsea, New York, 1977.

22. R. Gilmer, Multiplicative Ideal Theory, Marcel Dekker, New York, 1972.

23. R. Gilmer and R. Heitmann, On Pic(R[X]) for R seminormal, J. Pure Appl. Algebra 16 (1980), 251-257.

24. W. Green and E. Kamen, Stabilizability of linear systems over a commutative normed algebra with applications to spatially distributed and parameter-dependent systems, SIAM J. Contr. Opt. 23 (1985), 1-18.

25. M. Hautus and E. Sontag, New results on pole-shifting for parametrized families of systems, to appear.

26. M. Henriksen, On the prime ideals of the ring of entire functions, Pacific J. Math. 3 (1953), 711-720.

27. I. Herstein, Topics in Algebra, Blaisdell, New York, 1964.

28. M. Heymann, Comments on pole assignment in multi-input controllable linear systems, IEEE Trans. Automatic Control AC-13 (1968), 748-749.

29. B. Ho, An effective construction of realizations from input/output descriptions, doctoral dissertation, Stanford Univerisity, 1966.

30. R. Kalman, Mathematical description of linear dynamical systems, SIAM J. Control 1 (1963), 152-192.

31. R. Kalman, P. Falb, and M. Arbib, Topics in Mathematical System Theory, McGraw-Hill, New York, 1969.

32. E. Kamen, On an algebraic theory of systems defined by convolution operators, Math. System Theory 9 (1975), 57-74.

33. E. Kamen, Lectures on algebraic system theory: linear systems over rings, N.A.S.A. Contractor Report 3016 (1978).

34. E. Kamen, New results in realization theory for linear time-varying analytic systems, IEEE Trans. Aut. Control, AC-24 (1979), 866-878.

35. E. Kamen and P. Khargonekar, A transfer function approach to linear time-varying systems, Proc. IEEE Conf. Dec. and Control. Orlando, Dec. 1982, 152-157.

36. I. Kaplansky, Elementary divisors and modules, Trans. Amer. Math. Soc. 66 (1949), 464-491.

37. I. Kaplansky, Infinite Abelian Groups, revised edition, University of Michigan Press, Ann Arbor, 1969.

38. I. Kaplansky, Fields and Rings, University of Chicago Press, Chicago, 1969.

39. I. Kaplansky, Commutative Rings, University of Chicago Press, Chicago, 1974.

40. M. Laplaza, Some properties of the ring of entire functions, unpublished manuscript.

41. N. McCoy, Rings and Ideals, Mathematical Association of America, 1948.

42. B. McDonald, Linear Algebra over Commutative Rings, Marcel Dekker, New York, 1984.

43. A. Morse, Ring models for delay differential systems, Automatica 12 (1976), 529-531.

44. J. Mott, Convex directed subgroups of a group of divisibility, Can. J. Math. 26 (1974), 532-542.

45. C. Naudé and G. Naudé, Comments on pole assignability over rings, to appear.

46. G. Naudé and C. Nolte, A survey of the realization and duality theories of linear systems over rings, Quaestiones Mathematicae 5 (1982), 135-164.

47. V. Popov, Hyperstability of Control Systems, Springer-Verlag, Berlin, 1973. This is a translation and revision of Hiperstabilitatea Sistemelar Automatae, Editura Academiei R.S.R., Bucuresti, 1966.

48. S. Power, Hankel operators on Hilbert space, Bull. London Math. Soc. 12 (1980), 422-442.

49. D. Richman, A new proof of a result about Hankel operators, Integral Equations Operator Theory 5 (1982), 892-900.

50. C. Rickart, General Theory of Banach Algebras, Van Nostrand. New York, 1960.

51. J. Rotman, An Introduction to Homological Algebra, Academic Press, 1979.

52. Y. Rouchaleau, Linear, discrete time, finite dimensional dynamical systems over some classes of commutative rings, doctoral dissertation, Stanford University, 1972.

53. Y. Rouchaleau and E. Sontag, On the existence of minimal realizations of linear dynamical systems over noetherian integral domains, J. Comput. System Sci. 18 (1979), 65-75.

54. Y. Rouchaleau, B. Wyman, and R. Kalman, Algebraic structure of a linear dynamical systems. III. Realization theory over a commutative ring, Proc. Nat. Acad. Sci. (USA) 69 (1972), 3404-3406.

55. W. Rudin, Real and Complex Analysis, second edition, McGraw-Hill, New York, 1974.

56. D. Russell, Mathematics of Finite-dimensional Control Systems, Marcel Dekker, New York, 1979.

57. L. Silverman, Realizations of linear dynamical systems, IEEE Trans. Aut. Control, AC-16 (1971), 554-567.

58. L. Silverman and R. Bucy, Generalizations of a theorem of Dolezal, Math. Sys. Theory 4 (1970), 334-339.

59. E. Sontag, Linear systems over commutative rings: A survey, Ric di Automatica 7 (1976), 1-34.

60. E. Sontag, Parametric stabilization is easy, Systems and Control Letters 4 (1984), 181-188.

61. E. Sontag, An introduction to the stabilization problem for parametrized families of linear systems, AMS Summer Conference on Linear Algebra and its Applications to System Theory, Brunswick, Maine, August, 1984.

62. A. Tannenbaum, On pole assignability over polynomial rings, Systems and Control Letters 2 (1982), 13-16.

63. A. Tannenbaum, Polynomial rings over arbitrary fields in two or more variables are not pole assignable, Systems and Control Letters 2 (1982), 222-224.

64. R. Wiegand and S. Wiegand, Finitely generated modules over Bézout rings, Pac. J. Math. 58 (1975), 655-664.

65. W. Wonham, On pole assignment in multi-input controllable linear systems, IEEE Trans. Automatic Control AC-12 (1967), 660-665.

66. W. Wonham, Linear Multivariable Control: A Geometric Approach, Springer-Verlag, New York, 1979.

67. D. Youla and P. Tissi, n-port synthesis via reactance extraction. Part I, IEEE Intern. Convention Record, 1966.

68. O. Zariski and P. Samuel, Commutative Algebra, Volume I, Van Nostrand, Princeton, 1958.

Index

about the book . . .

Here is the first up-to-date, self-contained work to elucidate the statements and proofs of the fundamental results of linear systems over commutative rings. Emphasizing algebraic techniques, *Linear Systems Over Commutative Rings* synthesizes in a unified manner literature from various mathematical, engineering, and control theory journals.

Linear Systems Over Commutative Rings provides an overview of the relevant commutative ring theory . . . includes a general formulation of Sontag's recent results on stabilization of linear systems (unifying several results concerning parametric stabilization) . . . relates projective modules, Bézout domains, and the pole assignability property . . . organizes and clarifies results on Hankel matrices and Silverman's formulas . . . and describes results on rationality and descending realizability from the quotient field.

The book also incorporates a basic introduction to commutative algebra, as well as end-of-chapter exercises, and bibliographic data. It is a valuable resource for algebraists, systems and control theorists, and graduate students in algebra, systems theory, or control theory.

about the authors . . .

JAMES W. BREWER is Professor of Mathematics at the University of Kansas, Lawrence, where he has taught since 1970. He is the author of *Formal Power Series Over Commutative Rings*, the co-editor of *Emmy Noether: A Tribute to Her Life and Work* (both titles, Marcel Dekker, Inc.), and the author of many articles on commutative algebra. Dr. Brewer received the B.A. (1964) and Ph.D. (1968) degrees in mathematics from Florida State University. He is a member of the American Mathematical Society.

JOHN W. BUNCE is Professor of Mathematics at the University of Kansas, Lawrence, where he has been affiliated since 1970. He received the B.A. (1965) in mathematics from the University of Colorado, Boulder, and Ph.D. (1969) from Tulane University. Dr. Bunce's mathematical specialty is operator theory and operator algebras. He is a member of the American Mathematical Society.

F. S. VAN VLECK is Professor of Mathematics at the University of Kansas, Lawrence. He received the B.S. (1956) and M.A. (1957) degrees in mathematics from the University of Nebraska, and Ph.D. (1960) degree from the University of Minnesota. Dr. Van Vleck's areas of interest are ordinary differential equations, control theory, multiple-valued functions, selection problems, and optimization theory. He is a member of the American Mathematical Society and the Society for Industrial and Applied Mathematics.

Printed in the United States of America

ISBN: 0–8247–7559–7

marcel dekker, inc./new york · basel